Revealing Media Bias in News Articles

Felix Hamborg

Revealing Media Bias in News Articles

NLP Techniques for Automated Frame Analysis

Felix Hamborg
Department of Computer Science
Humboldt University of Berlin
Berlin, Germany

ISBN 978-3-031-17695-1   ISBN 978-3-031-17693-7   (eBook)
https://doi.org/10.1007/978-3-031-17693-7

This Springer imprint is published by the registered company Springer Nature Switzerland AG
The registered company address is: Gewerbestrasse 11, 6330 Cham, Switzerland

Für meine Eltern, Christine und Michael,
und meinen Bruder, Mario. Danke für alles!

Preface

News articles serve as a highly relevant source for individuals to inform themselves on current topics and salient political issues. How the news covers an issue decisively affects public opinion and our collective decision-making. Albeit the news is meant to not only communicate "objective facts" but also to assess events and their implications, biased coverage can be problematic. Especially when news consumers are not aware of the often subtle yet powerful slants present in the news, or when coverage is systematically slanted to alter public opinion, *media bias* poses a severe problem to society.

Empowering newsreaders to critically assess the news is an essential means to face the issues caused by media bias. On the one hand, non-technical means such as media literacy practices or analysis approaches devised in political science are highly effective. However, they often come with immense efforts, such as researching and contrasting relevant news articles. Ultimately, this effort can represent an insurmountable barrier for these manual techniques to be applied during daily news consumption. On the other hand, automated data analysis methods are available and could enable timely bias analysis. However, automated approaches largely neglect the sophisticated models and analysis approaches devised in decade-long bias research in the social sciences. Compared to them, the automated approaches often yield superficial or inconclusive results.

To enable effective and efficient bias identification, the thesis at hand proposes an interdisciplinary approach to reveal biases in English news articles reporting on a given political event. Therefore, the approach identifies the coverage's different perspectives on the event. The approach's so-called person-oriented frames represent how articles portray the persons involved in the event. In contrast to prior automated approaches, the identified frames are meaningful and substantially present in the news coverage. In particular, this thesis makes the following research contributions.

The thesis presents the first interdisciplinary literature review on approaches for analyzing media bias, thereby contrasting studies and models from the social sciences with automated approaches such as devised in computer science. A key finding is that research in either discipline could benefit from integrating the other's expertise and methods. To facilitate such interdisciplinary research, the thesis establishes a shared conceptual understanding by mapping the state of the art from the social sciences to a framework that automated approaches can target.

To address the weaknesses of prior work, the thesis then proposes *person-oriented framing analysis (PFA)*. The approach integrates methodology that has been applied in practice in social science research and identifies specific in-text means of narrowly defined bias forms. In contrast to prior automated approaches, which treat bias rather as a holistic, vague concept, PFA detects article groups representing meaningful frames. Such frames could previously be only identified through manual content analysis or expert knowledge on the analyzed topic.

Afterward, the thesis proposes methods for the PFA approach, investigates their suitability concerning PFA, and evaluates their technical effectiveness. For example, the thesis introduces the first method to classify sentiment in news articles. The thesis also lays out essential preparatory work for other tasks. For example, a method is proposed that resolves highly event-specific coreferences, which may even be of contradictory meanings in other contexts, such as "freedom fighters" and "terrorists."

To demonstrate the effectiveness of the PFA approach, a prototype system to reveal person-oriented framing in event coverage is presented and evaluated. The results of a user study ($n = 160$) demonstrate the effectiveness of the interdisciplinary approach devised in this thesis. In the study's single-blind setting, the PFA approach is most effective in increasing respondents' bias-awareness. Moreover, the study results confirm the findings of the literature review. They suggest that prior bias identification and communication approaches identify biases that are technically significant but often are meaningfully irrelevant. In practical terms, prior work facilitates the visibility of potential biases, whereas the PFA approach identifies meaningful biases indeed present in the coverage.

This thesis is motivated by my vision to mitigate media bias's severely adverse effects on societies. Outside the academic context, this vision entails that popular news aggregators and news apps will integrate effective approaches for bias identification, such as PFA, to help news readers critically assess news coverage in a practical, effortless way during daily news consumption.

Berlin, Germany   Felix Hamborg
April 2022

Acknowledgments

It is with deep gratitude that I am writing these lines—deep gratitude for the fruitful collaborations and generous support of numerous individual persons, teams, and institutions, without whom this thesis would not have been possible.

I am especially grateful to my doctoral advisor, Bela Gipp, for supporting me in all my research endeavors. I am very thankful for your inspiration, which has contributed a lot to become the researcher I am today. Likewise, I am very grateful to Karsten Donnay for his consistent support, critical comments, and valuable ideas. I have benefited a lot from your interdisciplinary perspective on my research, and so has this thesis. I also sincerely thank Falk Schreiber and Andreas Spitz for serving as members of my doctoral committee.

I especially acknowledge my research group at Newsalyze and Textalysis: Anastasia Zhukova, Timo Spinde, Moritz Bock, Kim Heinser, and Franziska Weeber. I am very grateful for our close collaborations, and I especially thank Anastasia for all her support. I sincerely thank also the other team members who contributed to the project's progress: Maximilian Kutzner, Jan Dix, Marc Luettecke, and Camille Lothe. It is a great pleasure to work with you.

I feel honored for having had the opportunity to advise students at the Universities of Konstanz, Freiburg, and Zurich, and the National Institute of Informatics (NII) in Tokyo. I thank the students who contributed to the news-please and Giveme5W1H projects, especially Moritz Bock, Sören Lachnit, and Benedikt Kromer. Working with students in lectures and supervising bachelor's and master's projects was and is a rewarding experience from which I have also learned so much.

I extend my sincere appreciation to all the passionate and inspiring colleagues from all over the globe with whom I had the distinct pleasure of working over the years. I particularly thank Norman Meuschke for his advice and critical comments, especially when I started as a doctoral researcher. I also thank Moritz Schubotz, Vincent Stange, Terry Ruas, and my other colleagues who supported me on so many occasions. I thank Stefan Gerlach for his support when I was using the computing cluster in Konstanz. I thank Christina Zuber for her critical comments and viewing my research from a political science perspective. I thank Christian Graf and Alexander Lang for their essential input back when I was a bachelor's student.

I am very thankful to Akiko Aizawa from the NII in Tokyo, Shivakumar Vaithyanathan and Berthold Reinwald from the IBM Almaden Research Center in San Jose, and Ming Hao from the HP Research Lab in Palo Alto. By inviting me to complete research stays in Tokyo and the Silicon Valley, you enabled invaluable experiences, both professionally and personally. A special thank you goes out to my NII friend Enguerrand Gentet. I also thank Adam Jatowt, Edi Rasmussen, Kentaro Toyama, Megan Fin, Sudipta Kar, and the other professors, postdocs, and doctoral researchers I got to know at several venues and doctoral workshops, especially at ACL, iConference, and JCDL. Your critical comments and fresh thoughts helped me to sharpen my doctoral research. I also thank my colleagues at the Committee for the Doctoral Fund at the University of Konstanz. Viewing one's own research discipline and projects from "outside the box" has always been a rewarding experience and source of inspiration.

I gratefully acknowledge the organizations that provided my colleagues and me with the funding to realize our research. I especially thank the Carl Zeiss Foundation for honoring me with a doctoral scholarship. Likewise, I thank the Heidelberg Academy of Sciences and Humanities for funding our media bias research. I thank the ACL, ACM, SIGIR, and the German Academic Exchange Service (DAAD) for their conference travel grants. I am also grateful for the Erasmus Plus Programme by the European Commission, and the MOU Grant Program by the NII for their generous support that allowed me to complete multiple research stays abroad. Lastly, I express my gratitude to the Heidelberg Academy of Sciences and Humanities and the University of Konstanz. They enabled my dissertation to be released as an open-access book.

I am extremely grateful to my friends. Special thanks go out to Philip Ehret, Christopher Fromm, Dimitri Gärtner, Dominik Grzelak, Jan Kiethe, Artem Kinsvater, Florian König, Tobias Leicher, Roman Matt, Marie Niedermeier, Alex Schaich, Philipp Scharpf, Alexander Schönhals, Sebastian Weigel, and Oliver Wiedemann. You rock!

Finally, and most importantly, I am very thankful to my family. Thank you for being there. I am incredibly thankful to my loving parents, Christine and Michael, and my awesome brother, Mario.

Thank you all!

Contents

Chapter 1
Introduction

Abstract This chapter motivates and summarizes my research to reveal bias in news articles. Section 1.1 highlights the issues caused by slanted news coverage. The importance of these issues cannot be overestimated as systematically biased coverage has already exerted influence on societal decisions, such as in democratic elections. Section 1.2 then briefly summarizes the research gap regarding the identification and communication of media bias. I take the issues caused by bias and the research gap in revealing bias as a motivation to define my research question and derive respective research tasks in Sect. 1.3. Lastly, Sect. 1.4 summarizes the contributions of this thesis and gives an overview of its structure and the integrated publications.

1.1 Problem

News articles serve as a highly relevant source for information on current events and salient political issues [61, 249, 365]. How an event or issue is covered in the news can decisively impact public debates and affect our collective decision-making. What if news as an essential source of information is biased? This is a highly relevant question since news coverage is very likely to be all but strictly "neutral." News is meant to put facts into context and assess events' implications. It is thus expected or even desirable for news to be biased.

News not being objective is not problematic per se [384]—as long as news consumers among the general public are aware of present biases [64]. Often, this is not the case. Studies have found severely harmful effects of slanted news coverage, e.g., on collective decision-making, public opinion, and policy decisions, such as in democratic elections [237, 259]. Thus, empowering news consumers to assess coverage critically is essential to address the issues caused by media bias. Media literacy is an appropriate means for critical news assessment and balanced interaction with the media. However, while such non-technical means can be highly effective, they require high effort, e.g., to research an event's articles and contrast their coverage. The high effort may represent an insurmountable barrier, preventing critical assessment in daily news consumption.

© The Author(s) 2023
F. Hamborg, *Revealing Media Bias in News Articles*,
https://doi.org/10.1007/978-3-031-17693-7_1

Table 1.1 Two news excerpts with different, almost opposing perspectives (also called *frames*) on the same event due to word choice and fact selection. Adapted from [363]

Publisher	Excerpt
New York Times	Iraqi fighter jets threatened two American U-2 surveillance planes, forcing them to abort their mission and to return.
USA Today	U.N. arms inspectors said they had withdrawn two U-2 reconnaissance planes over Iraq for safety reasons

To give a practical idea of media bias,[1] Table 1.1 shows excerpts of two related articles from 2003 that illustrate the subtle influence of how individual news events are reported. While both excerpts describe the same event, they portray different perspectives, commonly referred to as *(political) frames* [79].[2] Differences in the descriptions arise from including or excluding "facts" and using different words to refer to the actors, actions, objects, and reasons. In Table 1.1, *The New York Times* framed the Iraqi military as an aggressor threatening (peaceful) surveillance planes, while *USA Today* omitted the existence of Iraqi fighter jets and vaguely justified the withdrawal of the planes "for safety reasons." Beyond this example, the media used different overall frames for their coverage of the war itself: while Western media reported on the "War in Iraq," Iraqi news referred to it as the "War on Iraq." In sum, in the example depicted in Table 1.1, framing is achieved primarily by two forms of bias: *word choice* and *fact selection*.[3]

Adding to the complexity of bias and the diversity of its forms and resulting framing, the perception of topics can be altered through various means besides content, language, or generally text [14, 69, 146, 147, 276]. Other means, such as image selection, can similarly affect how news consumers perceive an event. For example, news outlets can choose different photos in their articles that show the same event and persons but in a different context or overall mood. In turn, readers' perception of the topic likely differs strongly depending on which article they read or picture they viewed.

While differences such as in word choice and selection of facts or photos are easy to notice when contrasting such opposing examples, spotting the resulting slants is very difficult or nearly impossible during daily news consumption. For example, newsreaders typically rely on only one or few similarly slanted news sources. Even if they are willing and trained to critically assess the news, researching

---

[1] Refer to Sect. 3.2, 3.2, for the definition of media bias used in this thesis. An overview of the various definitions used in prior research can be found in Sect. 2.2.1, 2.2.1.

[2] The term *(political) framing* as defined by Entman is different from Fillmore's *semantic frames* [87] in that political frames determine what a causal agent does with which benefit and cost. In contrast, semantic frames are defined to map words to their meaning in the context of fine-grained events.

[3] Both terms are used here for simplicity. The thesis will later introduce and refer to these terms using their established terms, i.e., "word choice and labeling" and "commission and omission of information" (Sect. 2.2.3).

and contrasting articles and facts causes strenuous effort. The high effort may prevent newsreaders from applying such effective yet cumbersome means routinely. Framing may then influence how we perceive specific information and assess events.

The potential negative effects of systematically biased media coverage, especially on policy issues, are manifold and can hardly be overestimated (cf. [24]). For example, a 2003 survey analyzed differences in news coverage on the Iraq War and corresponding perceptions of news consumers [71]. Fox News viewers were most misinformed: over 40% thought that weapons of mass destruction had been found in Iraq, which had been used by the US government as a justification for the war. Another study found a strong influence of news coverage on asylum policy decisions—even stronger than the impact of cultural or economic factors [184].

The problem of slanted coverage is further amplified since most people only rely on a few news sources [128], and news outlets are influenced by other media [379], often few central news agencies [33]. This is compounded by the fact that only a few corporations control large parts of the media landscape in many countries. In Germany, for example, only five corporations control more than half of the media [189], and in the USA, only six corporations control 90% [40, 318]. Further adding to the severity of media bias are recent trends in news production and consumption. More news is spread and consumed on other channels where news authors might more often disregard journalistic standards. Examples include social media channels, such as Facebook and Twitter, and alternative news portals, such as Breitbart.

In extreme cases, "fake news" may intentionally present entirely fabricated facts to manipulate public opinion toward a given topic, e.g., during the US presidential elections of 2016 [219]. While fake news is not systematically different from other types of biased news coverage, it represents the end of the spectrum insofar as biased news coverage may give way to gross distortion of facts or outright factually incorrect information. As in the example of Fox News viewers cited previously, the general population then ultimately no longer holds the same views on whether or not certain events have actually transpired. In the remainder of this thesis, the term media bias also entails the extreme but methodologically identical biases and their forms as they occur in fake news.

In sum, systematically biased coverage is a highly relevant and current issue. Biased coverage can decisively alter public opinion and poses severe societal challenges [9, 190, 237]. Empowering news consumers to critically assess coverage is essential, especially on policy issues. Albeit media literacy represents an effective means to a more balanced interaction with the media, it causes strenuous effort. Ultimately, this high effort prevents critical assessment of media during daily news consumption.

1.2   Research Gap

Enabling the comparison of substantially different perspective present in news coverage, such as shown in Table 1.1, would facilitate bias-sensitive consumption and assessment of the news. However, identifying such perspectives is currently only possible with strenuous effort using manual techniques since automated approaches cannot reliably detect them.

In computer science and related fields, media bias is a rather young research topic. Albeit technically more advanced, automated approaches tend to employ simpler models and methodology compared to manual bias analysis. Compared to the opposing views in Table 1.1, automated approaches find perspectives in event coverage that are technically different but often do not represent frames, i.e., meaningfully different perspectives. Ultimately, the approaches cannot enable news consumers to assess the news since the perspectives they identify critically may often be inconclusive, incomplete, or superficial. One key reason for their mixed results is that current approaches analyze—or generally treat—bias as an only vaguely defined concept, such as

"subtle differences" [211],
"differences of coverage" [278],
"diverse opinions" [251], or
"topic diversity" [252].

The shortcomings of automated approaches and their non-optimal results become apparent when comparing the approaches to research in the social sciences. There, decade-long research on media bias has resulted in comprehensive models to describe individual bias forms and effective methods to analyze them. Using established tools such as content or frame analysis, researchers in the social sciences can detect and quantify powerful and difficult-to-detect bias forms [155, 368]. For example, the data-driven analyses determine substantial frames by identifying in-text means (also called *framing devices*) from which these frames emerge. However, such analyses are conducted mostly manually, require much expertise, cause high cost, and can only be conducted for few topics in the past [155, 260].

In sum, critically assessing news coverage on policy issues is a crucial means to mitigate media bias's adverse effects. However, while reliable techniques for bias and frame identification are available, they cannot be used during daily consumption due to their high effort and required expertise. In contrast, scalable methods for automated data analysis, such as in natural language processing, are available. Current automated approaches to reveal biases, however, predominantly suffer from superficial results, especially when compared to social science research results. In our view, only an interdisciplinary approach can support newsreaders in critically assessing coverage during daily news consumption.

1.3 Research Question

I take the previously identified research gap as a motivation to define the following research question for my doctoral research:

How can an automated approach identify relevant frames in news articles reporting on a political event and then communicate the identified frames to non-expert news consumers to effectively reveal biases?

I intentionally define the research question rather openly to better reflect—compared to prior work in the social sciences—the recency and the relatively young state of the art in computer science. In the course of this thesis, I use the findings of the first interdisciplinary literature review on media bias (Chap. 2) to narrow down the research question to a specific research objective (Sect. 3.3.3). I then propose *person-oriented framing analysis (PFA)* to tackle the research objective (Sect. 3.4). In the evaluation of the approach using a prototype (Chap. 6) and the conclusion of this thesis (Chap. 7), I discuss the suitability of the conducted research not only concerning the specific research objective but also in the context of the broader research question.

Albeit the research summarized in this thesis focuses on news articles, the PFA approach can conceptually be applied to any news domain, source, and genre that focuses on persons and adheres to grammar and other linguistic rules. This includes alternative news channels, such as Breitbart, which I also include in the evaluation (Chap. 6). Albeit the PFA approach is conceptually language independent, I develop methods for bias analysis in English news articles. By using and devising methods for natural language understanding of the English language, I can demonstrate the best possible performance of the PFA approach.

To address the research question, I define the following research tasks:

RT 1 Identify the strengths and weaknesses of manual and automated methods used to identify and communicate media bias and its forms.

RT 2 Devise a bias identification approach that addresses the identified weaknesses of current bias identification approaches.

RT 3 Develop methods for the devised approach and evaluate their technical performance.

RT 4 Implement a prototype of a bias identification and communication system that employs the developed methods to reveal biases in real-world news coverage to non-expert news consumers.

RT 5 Evaluate the approach's effectiveness in revealing biases by testing the implemented prototype in a user study.

1.4   Thesis

This section gives an overview of the thesis at hand. Section 1.4.1 presents the structure of the thesis and its scientific contributions. Afterward, Sect. 1.4.2 introduces the peer-reviewed publications that this thesis summarizes and states how they are cited.

1.4.1   Structure and Scientific Contributions

Reading this thesis and its chapters in the provided order gives, in my opinion, the most intuitive access to the research summarized in this thesis. At the same time, each chapter is written to be understood without reading the other chapters first. Summaries of information presented in other chapters serve the purpose of good readability of individual chapters without readers having to follow cross-references often. Of course, in addition to these summaries, readers are provided with cross-references to the respective parts of the thesis where they can find more detailed information.

Chapter 1 presents a few of the severe problems caused by slanted news coverage and identifies the research gap that motivated the research described in this thesis. The chapter also introduces the research question that guided the summarized research.

Chapter 2 discusses manual analysis concepts and exemplary studies from the social sciences and automated approaches, mostly from computer science, to analyze and reveal media bias. Either of the disciplines uses distinctive terminology, and each has fundamentally different objectives and approaches. Thus, the chapter first establishes a shared conceptual understanding by mapping the state of the art from the social sciences to a framework that computer science approaches can target. In sum, the chapter identifies the strengths and weaknesses of current approaches for identifying and revealing media bias by presenting the **first interdisciplinary literature review** on the topic.

Addressed research task: RT 1
The publications summarized in Chap. 2 are [123, 126]

Chapter 3 discusses the solution design space to address the identified research gap. Then, the chapter devises **person-oriented framing analysis (PFA)**, our approach to identify substantial frames and to reveal slanted news coverage. PFA aims to detect groups of articles that report similarly on an event, i.e., that frame the event similarly, by determining how each article portrays the persons involved in the event.

Addressed research task: RT 2
The publications summarized in Chap. 3 are [134, 136]
Further publications relevant for the research described in this chapter, e.g., which report on earlier or preliminary results leading to the results described in the thesis, are [123, 137, 138]

Chapter 4 introduces target concept analysis (TCA), the first component of PFA after natural language preprocessing. Target concept analysis seeks to identify phrases that may be subject to specific biases in a set of news articles reporting on an event. Among others, the chapter introduces the **first method for context-driven cross-document coreference resolution**. In contrast to prior work, the method is capable of resolving highly topic- and event-specific coreferences that may even be antonyms in general, such as "coalition forces" and "invading forces."

Addressed research task: RT 3
The publications summarized in Chap. 4 are [124, 130]
Further publications relevant for the research described in this chapter are [131–133]

Chapter 5 presents frame identification, the second component of PFA. Conceptually, this component seeks to identify the person-oriented framing of individual articles. To approximate person-oriented framing, the method determines how individual sentences portray persons involved in the analyzed news event. Most importantly, the chapter introduces the **first large-scale dataset and a novel model for target-dependent sentiment classification (TSC) in the news domain**.

Addressed research task: RT 3
The publications summarized in Chap. 5 are [125, 127, 130]
A further publication, which reports on earlier results leading to the results described in the thesis, is [131]

Chapter 6 introduces Newsalyze, our **prototype system to reveal biases** to non-expert news consumers by using the PFA approach. The chapter first devises visualizations aimed to be intuitive and easy-to-use. The prototype system then integrates the visualizations and the methods devised in the previous chapters. In a large-scale user study, the PFA approach effectively increases bias-awareness in study participants. The prototype reveals substantial biases present in the news coverage, which in part could previously only be identified through manual frame analysis.

> Addressed research tasks: RT 4 and RT 5
> The publications summarized in Chap. 6 are [134, 137]

Chapter 7 summarizes the thesis and discusses the strengths and weaknesses of our research to derive ideas for future research on media bias.

Key Contributions

In sum, this thesis makes the following key contributions:

1. It presents the first interdisciplinary literature review on media bias combining expertise from computer science, the social sciences, and other disciplines relevant to analyzing media bias.
2. It proposes person-oriented framing analysis, an approach to identify and reveal meaningful frames, rather than only facilitating the visibility of potential perspectives.
3. It proposes a novel task named context-driven cross-document coreference resolution. This task aims to identify and resolve highly context-dependent coreferences as they occur frequently in slanted coverage. The thesis devises a dataset and method for this novel task.
4. It devises the first dataset and method for target-dependent sentiment classification (TSC) on news articles.
5. It introduces a prototype system including bias-sensitive visualizations to reveal media bias to non-expert news consumers by highlighting news articles' framing.
6. It presents the results of a user study that approximates real-world news consumption to demonstrate the approach's effectiveness concerning the change of bias-awareness in respondents.

One key finding of this thesis is that the devised methods and in particular the overview visualizations most effectively help news consumers to become aware of biases present in the news. Moreover, the results suggest that the developed system, Newsalyze, is the first to identify meaningful framing in person-centric coverage. Prior, such framing could only be identified using manual analyses or when already having an extensive understanding of a news topic.

Side Contributions

Individual chapters and sections describe further contributions required for my doctoral research, including approaches for news crawling and information extraction (Sect. 3.5) and main event retrieval from news articles (Sect. 4.2). Another side contribution is the exploratory research on automatically determining how news articles portray individual persons using so-called frame properties that represent topic-independent framing categories (Sect. 5.2).

Table 1.2 Overview of the core publications describing the research summarized or used in this thesis. The publication types C, J, and W represent conferences, journals, and workshops, respectively

Chapter	Venue	Year	Type	Length	Reference
1, 2	IJDL	2019	J	full	[126]
1, 2, 3	ACL	2020	W	full	[123]
3, 6	RecSys	2021	W	short	[134]
3	JCDL	2020	C	poster	[138]
3	ISI	2017	C	short	[136]
4	JCDL	2019	C	full	[130]
4	iConf.	2019	C	short	[131]
4	iConf.	2018	C	full	[133]
4	JCDL	2018	C	poster	[132]
4	RecSys	2019	W	full	[124]
5	iConf.	2021	C	short	[127]
5	EACL	2021	C	full	[125]
5, 6	JCDL	2020	C	poster	[96]
6	JCDL	2020	C	short	[332]
6, 3	JCDL	2021	C	full	[137]

1.4.2   Publications

To subject my research to the scrutiny of peer review, I have published all major contributions of this thesis in conference proceedings and journals. Four of the publications were honored an award or were nominated for one, two of which are directly relevant to this thesis ([131, 133]) and three of which I am the responsible first author ([129, 131, 133]). More information on the awarded publications can be found in Appendix A.5.

When writing the thesis, I aimed to achieve a trade-off between a well-readable dissertation (rewriting all my peer-reviewed publications) and a thesis following the strictest citation rules (quoting all sections related to a publication). All publications that are directly relevant to the thesis at hand are shown in Table 1.2. These are the origin of the text and other content I use in this thesis. The first column in Table 1.2 indicates which chapter is based on which publications. When using my own publications for this thesis, I copied the content of the publication and adapted words or larger parts, e.g., for consistent wording, to better fit into the overall structure of the thesis or to reflect recent literature and developments that happened since writing the original publication.

To acknowledge the fellow researchers with whom I published, collaborated, and discussed ideas, I will use "we" instead of "I" in the remainder of this thesis.

Chapter 2
Media Bias Analysis

Abstract This chapter provides the first interdisciplinary literature review on media bias analysis, thereby contrasting manual and automated analysis approaches. Decade-long research in political science and other social sciences has resulted in comprehensive models to describe media bias and effective methods to analyze it. In contrast, in computer science, computational linguistics, and related fields, media bias is a relatively young research topic. Despite many approaches being technically very advanced, we find that the automated approaches could often yield more substantial results by using knowledge from social science research on the topic.

2.1 Introduction

The Internet has increased the degree of self-determination in how people gather knowledge, shape their own views, and engage with topics of societal relevance [249]. Unrestricted access to unbiased information is crucial for forming a well-balanced understanding of current events. For many individuals, news articles are the primary source to attain such information. News articles thus play a central role in shaping personal and public opinion. Furthermore, news consumers rate news articles as having the highest quality and trustworthiness compared to other media formats, such as TV or radio broadcasts, or, more recently, social media [61, 249, 365]. However, media coverage often exhibits an internal bias, reflected in news articles and commonly referred to as *media bias*. Factors influencing this bias can include ownership or source of income of the media outlet or a specific political or ideological stance of the outlet and its audience [363].

The literature identifies numerous ways in which media coverage can manifest bias. For instance, journalists *select events, sources,* and from these sources the *information* they want to publish in a news article. This initial selection process introduces bias to the resulting news story. Journalists can also affect the reader's perception of a topic through *word choice*, e.g., if the author uses a word with a positive or a negative connotation to refer to an entity [116], or by varying the credibility ascribed to the source [14, 99, 266]. Finally, the *placement* and *size* of an

© The Author(s) 2023
F. Hamborg, *Revealing Media Bias in News Articles*,
https://doi.org/10.1007/978-3-031-17693-7_2

article within a newspaper or on a website determine how much attention the article will receive [37].

The impact of media bias, especially when implemented intentionally (see the review of bias definitions in Sect. 2.2.1), on shaping public opinion has been studied by numerous scholars [24]. Historically, major outlets exerted a strong influence on public opinion, e.g., in elections [219, 237, 259], or the social acceptance of tobacco consumption [9, 362]. The influence of media corporations has increased significantly in the past decades. In Germany, for example, only five corporations control more than half of the media [189], and in the USA, only six corporations control 90% [40, 318]. This naturally increases the risk of media coverage being intentionally biased [82, 342]. Also on *social media*, which typically reflects a broader range of opinions, people may still be subject to media bias [10, 15, 111], despite social media being characterized by more direct and frequent interaction between users, and hence presumably more exposure to different perspectives. Some argue that social media users are more likely to actively or passively isolate themselves in a "filter bubble" or "echo chamber" [352], i.e., only be surrounded by news and opinions close to their own. However, this isolation is not necessarily as absolute as often assumed, e.g., Barberá et al. [17] found noticeable isolation for political issues but not for others, such as reporting on accidents and disasters. Recent technological developments are another reason for topical isolation of social media consumers, which might lead to a general decrease in the diversity of news consumption. For instance, Facebook, the world's largest social network with more than three billion users [85], introduced *Trending Topics* in 2014, a news overview feature. There, users can discover current events by exclusively relying on Facebook. However, the consumption of news from only a single distributor amplifies the previously mentioned level of influence further: only a single company controls what is shown to news consumers.

The automated identification of media bias and the analysis of news articles in general have recently gained attention in computer science. A popular example are news aggregators, such as *Google News*, which give news readers a quick overview of a broad news landscape. Yet, established systems currently provide no support for showing the different perspectives contained in articles reporting on the same news event. Thus, most news aggregators ultimately tend to facilitate media bias [39, 375]. Recent research efforts aim to fill this gap and reduce the effects of such biases. However, the approaches suffer from practical limitations, such as being fine-tuned to only one news category or relying heavily on user input [252, 253, 276]. As we show in this chapter, an important reason for the comparably poor performance of the technically superior computer science methods for automatic identification of instances of media bias is that such approaches currently tend to not make full use of the knowledge and expertise on this topic from the social sciences.

This chapter is motivated by the question of how computer science approaches can contribute to identifying media bias and mitigating the negative bias effects by ultimately making available a more balanced coverage of events and societal issues to news consumers. We address this question by comparing and contrasting established research on the topic of media bias in the social sciences with technical approaches from computer science. This comparative review thus also serves as a guide for computer scientists to better benefit from already more established media bias research in the social sciences. Similarly, social scientists seeking to apply current automated approaches to their own media bias research will also benefit from this review.

The remainder of this chapter is structured as follows. In Sect. 2.2, we introduce the term media bias, highlight the effects of slanted news coverage, provide an understanding of how bias arises during the production of news, and introduce the most important approaches from the social sciences to analyze media bias. Then, each of the subsections in Sect. 2.3 focuses on a specific *form* of media bias, describes studies from the social sciences that analyze this form, and discusses methods from computer science that have been used or could be used to identify the specified form of bias automatically. In Sect. 2.4, we discuss the reliability and generalizability of the manual approaches from the social sciences and point out key issues to be considered when evaluating interdisciplinary research on media bias. Section 2.5 summarizes the key findings of our literature review. Section 2.6 demonstrates the key findings and research gap using a practical example. Lastly, Sect. 2.7 summarizes the findings of the chapter in the context of this thesis.

2.2   Media Bias

This section gives an overview of definitions of media bias as used in social science research on the topic or as employed by automated approaches (Sect. 2.2.1). Afterward, we describe the effects of biased news coverage (Sect. 2.2.2), develop a conceptual understanding of how media bias arises in the process of news production (Sect. 2.2.3), and briefly introduce the most important approaches from the social sciences to analyze bias in the media (Sect. 2.2.4).

2.2.1   Definitions

The study of biased news coverage has a long tradition in the social sciences going back at least to the 1950s [253]. In the classical definition of Williams, media bias must both be intentional, i.e., reflect a conscious act or choice, and be sustained, i.e., represent a systematic tendency rather than an isolated incident [382]. This definition sets the media bias that we consider apart from other sources of *unintentional* bias in news coverage. Sources of unintentional bias include the influence of *news values* [141] throughout the production of news [276] and later the news consumption by readers with different backgrounds [266]. Examples for news values include the geographic vicinity of a newsworthy event to the location of the news outlet and consumers or the effects of the general visibility or societal relevance of a specific topic [229].

Many other definitions of media bias and its specific forms exist, each depending on the particular context and research questions studied. Mullainathan and Shleifer define two high-level types of media bias concerned with the intention of news outlets when writing articles: *ideology* and *spin* [327]. Ideological bias is present if an outlet biases articles to promote a specific opinion on a topic. Spin bias is present if the outlet attempts to create a memorable story. Another definition of media bias that is commonly used distinguishes between three types: *coverage*, *gatekeeping*, and *statement* (cf. [64]). Coverage bias is concerned with the visibility of topics or entities, such as a person or country, in media coverage. Gatekeeping bias, also called selection bias or agenda bias, relates to which stories media outlets select or reject for reporting. Statement bias, also called presentation bias, is concerned with how articles choose to report on concepts. For example, in the US elections, a well-observed bias arises from *editorial slant* [75], in which the editorial position on a given presidential candidate affects the quantity and tone of a newspaper's coverage. Further forms and definitions of media bias can be found in the discussion by D'Alessio and Allen [64].

Even more definitions of media bias are found when considering research on automated bias analysis. Automated approaches tackle media bias, for example, as "subtle differences" [210], "differences of coverage" [278], "diverse opinions" [251], or "topic diversity" [252]. In sum, these definitions are rather superficial and vague, especially when compared to social science research.

To closely resemble how bias is analyzed in the social sciences, we follow in this literature review the traditional definition by Williams as mentioned previously [382]. To allow for an extensive overview of media bias literature, we also include studies that are not strictly concerned with intentional biases only. To address the different objectives of social science research on media bias and our thesis, we later provide a task-specific definition of media bias that we use in the methodology chapters of our thesis (Chap. 3). Specifically, classical research on media bias in the social sciences is concerned with investigating bias as systematic tendencies or patterns in news coverage on more extended time frames, e.g., to measure the influence of (biased) coverage on society or policy decisions. In contrast, our research question is concerned with biases in current coverage, e.g., to inform news consumers about such biases. Thus, to enable timely bias communication to news consumers, we explicitly allow for biases that may or may not have tendencies on larger time frames.

2.2.2   Effects of Biased News Consumption

Media bias has a strong impact on both individual and public perception of news events and thus impacts political decisions [24, 69, 97, 100, 159, 166, 399]. Despite the rise of social media, news articles published by well-established media outlets remain the primary source of information on current events (cf. [61, 249, 365]). Thus, if the reporting of a news outlet is biased, readers are prone to adopting similarly biased views. Today, the effects of biased coverage are amplified by social media, in which readers tend to "follow" only the news that conforms with their established views and beliefs [92, 117, 250, 254, 351]. On social media, news readers encounter an "echo chamber," where their internal biases are only reinforced. Furthermore, most news readers only consult a small subset of available news outlets [261, 262], as a result of information overload, language barriers, or their specific interests or habits.

Nearly all news consumers are affected by media bias [72, 190, 194, 237, 259], which may, for example, influence voters and, in turn, influence election outcomes [71, 75, 196, 237, 259]. Another effect of media bias is the polarization of public opinion [352], which complicates agreements on contentious topics. These negative effects have led some researchers to believe that media bias challenges the pillars of our democracy [166, 399]: if media outlets influence public opinion, is the observed public opinion really the "true" public opinion? For instance, a 2003 survey showed that there were significant differences in the presentation of information on US television channels [190]. Fox News viewers were most misinformed about the Iraq War. Over 40% of viewers believed that weapons of mass destruction were actually found in Iraq, which is the reason used by the US government to justify the war.

According to social science research, the three key ways in which media bias affects the perception of news are *priming*, *agenda setting*, and *framing* [75, 314]. Priming theory states that how news consumers tend to evaluate a topic is influenced by their (prior) perception of the specific issues that were portrayed in news on that topic. Agenda setting refers to the ability of news publishers to influence which topics are considered relevant by selectively reporting on topics of their choosing. News consumers' evaluation of topics is furthermore based on the perspectives portrayed in news articles, which are also known as *frames* [79]. Journalists use framing to present a topic from their perspective to "promote a particular interpretation" [80].

We illustrate the effect of framing using an example provided by Kahneman and Tversky [166]: Assume a scenario in which a population of 600 people is endangered by an outbreak of a virus. In a first survey, Kahneman and Tversky asked participants which option they would choose:

A. **200 people will be saved.**
B. 33% chance that 600 people will be saved. 66% chance that no one will be saved.

In the first survey, 72% of the participants chose A, and 26% chose B. Afterward, a second survey was conducted that objectively represents the exact same choices, but here the options to choose from were framed in terms of likely deaths rather than lives saved.

C. 400 people will die.
D. **33% chance that no one will die. 66% chance that 600 people will die.**

In this case, the preference of participants was reversed. 22% of the participants chose C, and 72% chose D. The results of the survey thus demonstrated that framing alone, that is, the way in which information is presented, has the ability to draw attention to either the negative or the positive aspects of an issue [166].

In summary, the effects of media bias are manifold and especially dangerous when individuals are unaware of the occurrence of bias. The recent concentration of the majority of mass media in the hands of a few corporations amplifies the potential impact of media bias of individual news outlets even further.

2.2.3   *Understanding Media Bias*

Understanding not only various forms of media bias but also at which stage in the news production process they can arise [276] is beneficial to devise methods and systems that help to reduce the impact of media bias on readers. We focus on a specific conceptualization of the news production process, depicted in Fig. 2.1, which models how media outlets turn events into news stories and how then readers consume the stories (cf. [14, 69, 146, 147, 276, 277]). The stages in the process map to the forms of bias described by Baker, Graham, and Kaminsky [14]. Since each stage of the process is distinctively defined, we find this conceptualization of the news production process and the included bias forms to be the most comprehensive model of media bias for the purpose of devising future research in computer science. In the following paragraphs, we exemplarily demonstrate the different forms of media bias within the news production and consumption process. In Sect. 2.3, we discuss each form in more detail. Note that while the process focuses on news articles, most of our discussion in Sect. 2.3 can be adapted to other media types, such as social media, blogs, or transcripts of newscasts.

Various parties can directly or indirectly, intentionally or structurally influence the news production process (refer to the motives underlying media bias shown in the orange rectangle in Fig. 2.1). News producers have their own *political* and *ideological views* [59]. These views extend through all levels of a news company, e.g., news outlets and their journalists typically have a slant toward a certain political direction [117]. Journalists might also introduce bias in a story if the change is supportive of their career [19]. In addition to these internal forces, external factors may also influence the news production cycle. News stories are often tailored for a current *target audience* of the news outlet [98, 117, 220], e.g., because readers switch to other news outlets if their current news source too often contradicts their own beliefs and views [92, 98, 250, 254, 351]. News producers may tailor news stories for their *advertisers and owners*, e.g., they might not report on a negative event involving one of their main advertisers or partnered companies [69, 103, 220]. Similarly, producers may bias news in favor of *governments* since they rely on them as a source of information [25, 65, 146].

In addition to these external factors, business reasons can also affect the resulting news story, e.g., investigative journalism is more expensive than copy-editing prepared press releases. Ultimately, most news producers are profit-oriented companies that may not claim the provision of bias-free information to their news

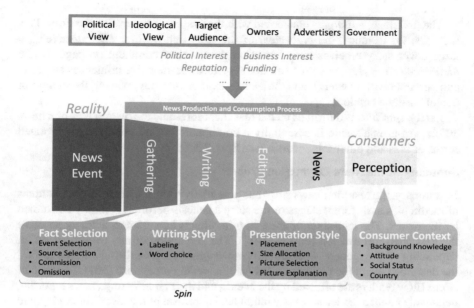

Fig. 2.1 The news production process is a model explaining how forms of bias emerge during the process of turning an event (happening in *reality*) into a news item (which is then perceived by news *consumers*). The orange part at the top represents internal and external factors that influence the production of a news item and its slants. The green parts at the bottom represent bias forms that can emerge during the three phases of the news production process. The "consumer context" label (far right) additionally shows factors influencing the perception of the described news event that are not related to media bias. Adapted from [276]

consumers as their main goal [281]; in fact, news consumers expect commentators to take positions on important issues and filter important from unimportant information (cf. [31, 81]).

All these factors influence the news production process at various stages (gray). In the first stage, *gathering*, journalists *select facts* from all the news events that happened. This stage starts with the *selection* of *events,* also named story selection. Naturally, not all events are relevant to a new outlet's target audience, or sensational stories might yield more sales [117]. Next, journalists need to *select sources*, e.g., press releases, other news articles, or studies, to be used when writing an article. Ultimately, the journalists must decide which information from the sources to be included and which to be excluded from the article to be written. This step is called *commission* or *omission* and likewise affects which perspective is taken on the event.

In the next phase, *writing*, journalists may use different *writings styles* to bias news. For instance, two forms defined in the production process are *labeling* (e.g., a person is labeled positively, "independent politician," whereas for the other party, no label or a negative label is used) and *word choice* (e.g., how the article refers to an entity, such as "coalition forces" vs. "invading forces").

The last stage, *editing*, is concerned with the *presentation style* of the story. This includes, for instance, the *placement* of the story and the *size allocation* (e.g., a large cover story receives more attention than a brief comment on page 3), the *picture selection* (e.g., usage of emotional pictures or their size influences attention and perception of an event), and the *picture explanation* (i.e., placing the picture in context using a caption).

Lastly, *spin bias* is a form of media bias that represents the overall bias of a news article. An article's spin is essentially a combination of all previously mentioned forms of bias and other minor forms (see Sect. 2.3.8).

Summary of the News Production Process

In summary, the resulting news story has potentially been subject to various sources of media bias at different stages of the story's genesis before it is finally consumed by the reader. The *consumer context*, in turn, affects how readers actually perceive the described information (cf. [16, 348]). The perception of any event will differ, depending on the readers' *background knowledge*, their preexisting *attitude* toward the described event (sometimes called *hostile media perception*) [367], their *social status* (how readers are affected by the event), and their *country* (e.g., news reporting negatively about a reader's country might lead to refusal of the discussed topic), and a range of other factors. Note, however, that "consumer context" is not a form of media bias and thus will be excluded from analysis in the remainder of this chapter.

Other Bias Models

Other models exist of how media bias arises, but their components can effectively be mapped to the news production and consumption process detailed previously. For instance, Entman defines a *communication process* that essentially mirrors all the same steps discussed in Fig. 2.1: (1) Communicators make intentional or unintentional decisions about the content of a text. (2) The text inherently contains different forms of media bias. (3) Receivers, i.e., news readers, draw conclusions based on the information and style presented in the text (which, however, may or may not reflect the text's perspective). (4) Receivers of a social group are additionally subject to *culture*, also known as a common set of perspectives [79].

Table 2.1 gives an overview of the previously described forms of media bias, where the "medium" column shows the medium that is the source of the specific form of bias and the column "target object" shows the items within the target medium that are affected.

2.2.4 Approaches in the Social Sciences to Analyze Media Bias

Researchers from the social sciences primarily conduct so-called content analyses to identify and quantify media bias in news coverage [64] or to, more generally, study patterns in communication. First, we briefly describe the concept and workflow of content analysis. Next, we describe the concept of *frame analysis*, which is a

Table 2.1 Overview of bias forms. The second column contains for each form of bias references to an exemplary study from the social sciences and the most relevant publications from computer science, if any

Name	Prior Work	Medium	Target Object	Phase	Explanation / Example
Event selection	[119, 34, 307]	News outlet	News article	Gathering	The news outlet rarely report on events criticizing the government.
Source selection	[117, 309, 5]	News article, text	Text, picture	Gathering	Inclusion of more sources that report on a certain perspective.
Commission and omission	[276, 98, 106]	News article, text	Text	Gathering	Facts that support or question a specific perspective are added to or omitted from the article.
Word choice and labeling	[116, 274, 28]	Text	Entity, action, attribute, etc.	Writing	Liberal vs. conservative, expert, independent; intervene vs. invade, clever vs. sneaky, refugee vs. immigrant
Story placement	[340]	News outlet	News article	Editing	Cover story receives more attention than a 3rd page story.
Size allocation	[307, 337]	News outlet	News article	Editing	A large story is likely to receive more attention than a small story.
Picture selection	[371, 317, 41]	News article, picture	Picture, entity, action, etc.	Editing	What does the picture show? For example fighting vs. a peace flag.
Picture explanation	[328]	Text	Picture caption	Editing	The caption puts the picture into context and may either support or criticize what is shown.
Spin	[277, 52, 1, 128]	News article & outlet	One or more news articles	All phases	The overall slant of the article, i.e., the result when the various types of news bias are combined.

specialized form of content analysis commonly used to study the presence of frames in news coverage [368]. Lastly, we introduce *meta-analysis*, in which researchers combine the findings from other studies and analyze general patterns across these studies [155].

2.2.4.1   Content Analysis

Content analysis quantifies media bias by identifying and characterizing its instances within news texts. In a content analysis, researchers first define one or more analysis questions or hypotheses. Researchers then gather the relevant news data, and coders (also called annotators) systematically read the news texts, annotating parts of the texts that indicate instances of media bias relevant to the analysis being performed. Afterward, the researchers use the annotated findings to accept or reject their hypotheses [228, 267].

In a *deductive* content analysis, researchers devise a *codebook* before coders read and annotate the texts [68, 227]. The codebook contains definitions, detailed rules, and examples of what should be annotated and in which way. Sometimes, researchers reuse existing codebooks, e.g., Papacharissi and de Fatima Oliveira [274] used annotation definitions from a previous study by Cappella and Jamieson [44] to create their codebook, and then they performed a deductive content analysis comparing news coverage on terrorism in the USA and the UK.

In an *inductive* content analysis, coders read the texts without specified instructions on how to code the text, only knowing the research question [117]. Since statistically sound conclusions can only be derived from the results of deductive content analyses [260], researchers conduct inductive content analyses mainly in early phases of their research, e.g., to verify the line of research or to find patterns in the data and devise a codebook [260, 368].

Usually, creating and refining the codebook is a time-intensive process, during which multiple analyses or tests using different iterations of a codebook are performed. A common criterion that must be satisfied before the final deductive analysis can be conducted is to achieve a sufficiently high inter-coder reliability (ICR) or inter-rater reliability (IRR) [195]. ICR, also called inter-coder agreement, inter-annotator reliability, or inter-annotator agreement, represents how often individual coders annotate the same parts of the documents with the same codes from the codebook. IRR represents this kind of agreement as well, but in a labeling task, e.g., with a fixed set of labels to choose from, rather than (also) annotating a phrase in a text. In some cases, these terms and tasks may overlap. In the remainder of this thesis, we will generally use the term ICR for annotation tasks where phrases have to be selected (and labeled), such as in a content analysis. We will use the term IRR for labeling tasks, e.g., where only one or more labels have to be selected but the phrase is given, such as in sentiment classification.

Social scientists distinguish between two types of content analyses: quantitative and qualitative [366]. A qualitative analysis seeks to find "all" instances of media bias, including subtle instances that require human interpretation of the text. In a quantitative analysis, researchers in the social sciences determine the frequency of specific words or phrases (usually as specified in a codebook). Additionally, researchers may subsume specific sets of words to represent so-called salient topics, roughly resembling frames (cf. [63]). Quantitative content analyses may also measure other, non-textual features of news articles, such as the number of articles published by a news outlet on a certain event or the size and placement of a story in a printed newspaper. These measurements are also called *volumetric measurements* [64].

Thus far, the majority of studies on media bias performed in the social sciences conduct qualitative content analyses because the findings tend to be more comprehensive. Quantitative analyses can be performed faster and can be partially automated, but are more likely to miss subtle forms of bias [316]. We discuss both qualitative and quantitative analyses for the individual bias forms in Sect. 2.3.

Content analysis software, generally also called *computer-assisted qualitative data analysis software (CAQDAS)*, supports analysts when performing content analyses [215]. Most tools support the manual annotation of findings for the analyzed news data or for other types of reports, such as police reports [267]. To reduce the large amount of texts that need to be reviewed, the software helps users find relevant text passages, e.g., by finding documents or text segments containing the words specified in the codebook or from a keyword list [336] so that the coder must review less texts manually. In addition, most software helps users find patterns in the documents, e.g., by analyzing the frequencies of terms, topic, or word co-occurrences [215].

2.2.4.2 Frame Analysis

Frame analysis (also called framing analysis) investigates how readers perceive the information in a news article [79]. This is done by broadly asking two questions: (1) *What* information is conveyed in the article? (2) *How* is that information conveyed? Both questions together define a *frame*. As described in Sect. 2.2.2, a frame is a selection of and emphasis on specific parts of an event.

To empirically determine the frames in news articles or other news texts, frame analysis is typically concerned with one or more of four dimensions [271]: syntactical, script, thematic, and rhetorical. The syntactical dimension includes patterns in the arrangement of words and, more broadly, information, e.g., descending order of salience in a story. The script dimension refers to characteristics similar to those of a story, i.e., a news article may have an introduction, climax, and end. The thematic dimension refers to which information is mentioned in a news text, e.g., which "facts," events, or sources are mentioned or quoted to strengthen the text's argument. Lastly, the rhetorical dimension entails how such information is presented, e.g., the word choice. Using these dimensions, researchers can systematically analyze and quantify the viewpoints of news texts.

Not all frame analyses focus on the text of news articles. For instance, DellaVigna and Kaplan [71] analyzed the gradual adoption of cable TV of Fox News between 1996 and 2000 to show that Fox News had a "significant impact" [71] on the presidential elections. Essentially, the study analyzed whether a district had already adopted the Fox News channel and what the election result was. The results revealed that the Republican Party had an increased vote share in those towns that had adopted Fox News.

2.2.4.3   Meta-Analysis

In a *meta-analysis*, researchers combine the results of multiple studies to derive further findings from them [155]. For example, in the analysis of event selection bias, a common question is which factors influence whether media organizations will choose to report on an event or not. McCarthy, McPhail, and Smith [229] performed a meta-analysis of the results of prior work suggesting that the main factors for media to report on a demonstration are the demonstration size and the previous media attention on the demonstration's topic.

2.2.5   *Summary*

News coverage has a strong impact on public opinion, i.e., what people think about (*agenda setting*), the context in which news is perceived (*priming*), or how topics are communicated (*framing*). Researchers from the social sciences have extensively studied such forms of media bias, i.e., the intentional, non-objective coverage of news events. The research has resulted in a broad literature on different forms and possible sources of media bias and their impact on (political) communication or opinion formation. In tandem, various well-established research methodologies, such as content analysis, frame analysis, and meta-analysis, have emerged in the social sciences.

The three forms of analysis discussed in Sect. 2.2.4 require significant manual effort and expertise [276], since those analyses require human interpretation of the texts and cannot be fully automated. For example, a quantitative content analysis might (semi-)automatically count words that have previously been manually defined in a codebook, but they would be unable to read for "meaning between the lines," which is why such methods continue to be considered less comprehensive than a qualitative analysis. However, the recent methodological progress in natural language processing (NLP) in computer science promises to help alleviate many of these concerns.

In the remainder of this chapter, we discuss the different forms of media bias defined by the news production and consumption process. The process we have laid out in detail previously is in our view the most suitable conceptual framework to map analysis workflows from the social sciences to computer science and thus helps us to discuss where and how computer scientists can make unique contributions to the study of media bias.

2.3 Manual and Automated Approaches to Identify Media Bias

This section is structured into nine subsections discussing all of the forms of media bias depicted in Table 2.1. In each subsection, we first introduce each form of bias and then provide an overview of the studies and techniques from the social sciences used to analyze that particular form. Subsequently, we describe methods and systems that have been proposed by computer science researchers to identify or analyze that specific form of media bias. Since media bias analysis is a rather young topic in computer science, often no or few methods have been specifically designed for that specific form of media bias, in which case, we describe the methods that could best be used to study the form of bias. Each subsection concludes with a summary of the main findings highlighting where and how computer science research can make a unique contribution to the study of media bias.

2.3.1 Event Selection

From the countless stream of events happening each day, only a small fraction can make it into the news. Event selection is a necessary task, yet it is also the first step to bias news coverage. The analysis of this form of media bias requires both an event-specific and a long-term observation of multiple news outlets. The main question guiding such an analysis is whether an outlet's coverage shows topical patterns, i.e., some topics are reported more or less in one as compared to another outlet, or which factors influence whether an outlet reports on an event or not.

To analyze event selection bias, at least two datasets are required. The first dataset consists of news articles from one or more outlets; the second is used as a ground truth or baseline, which ideally contains "all" events relevant to the analysis question. For the baseline dataset, researchers from the social sciences typically rely on sources that are considered to be the most objective, such as police reports [119]. After linking events across the datasets, a comparison enables researchers to deduce factors that influence whether a specific news outlet reports on a given event. For instance, several studies compare demonstrations mentioned in police reports with news coverage on those demonstrations [228, 229, 267]. During the manual content analyses, the researchers extracted the type of event, i.e., whether it was a rally, march, or protest, the issue the demonstration was about, and the number of participants. Two studies found that the number of participants and the issue of the event, e.g., protests against the legislative body [267], had a high impact on the frequency in news coverage [119].

Meta-analyses have also been used to analyze event selection bias, mainly by summarizing findings from other studies. For instance, D'Alessio and Allen found that the main factors influencing media reporting on demonstration are the demonstration size and the previous media attention on the demonstration's topic [64].

To our knowledge, only few automated approaches have been proposed that specifically aim to analyze event selection bias. Other than in social sciences studies, none of them compares news coverage with a baseline that is considered objective, but they compare the coverage of multiple outlets or other online news sources [34, 307]. In the following, we first describe these approaches in more detail, and then we also describe current methods and systems that could support the analysis of this form of bias.

Bourgeois, Rappaz, and Aberer [34] span a matrix over news sources and events extracted from GDELT [201], where the value of each cell in the matrix describes whether the source (row) reported on the event (column) [215]. They use matrix factorization (MF) to extract "latent factors," which influence whether a source reports on an event. Main factors found were the affiliation, ownership, and geographic proximity of two sources. Saez-Trumper, Castillo, and Lalmas [307] analyze relations between news sources and events. By analyzing the overlap between news sources' content, they find, for example, that news agencies, such as AP, publish most non-exclusive content—i.e., if news agencies report on an event, other news sources will likely also report on the event—and that news agencies are more likely to report on international events than other sources. Media type was also a relevant event selection factor. For example, magazine-type media, such as *The Economist*, are more likely to publish on events with high prominence, i.e., events that receive a lot of attention in the media.

Similar to the manual analyses performed in the social sciences, automated approaches need to (1) find or use articles relevant to the question being analyzed (we describe relevant techniques later in this subsection; see the paragraphs on news aggregation), (2) link articles to baseline data or other articles, and (3) compute statistics on the linked data.

In task (2), we have to distinguish whether one wants to compare articles to a baseline, or, technically said, across different media, or to other articles. Linking events from different media, e.g., news articles and tweets on the same events, has recently gained attention in computer science [307, 361]. However, to our knowledge, there are currently no *generic* methods to extract the required information from police reports or other, non-media databases, since the information that needs to be extracted depends on the particular question studied and the information structure and format differ greatly between these documents, e.g., police reports from different countries or states usually do not share common formats (cf. [206, 231]).

To link news articles reporting on the same event, various techniques can be used. *Event detection* extracts events from text documents. Since news articles are usually concerned with events, event detection is commonly used in news-related analyses. For instance, in order to group related articles, i.e., those reporting on the same event [164], one needs to first find events described in these articles. *Topic modeling* extracts semantic concepts, or topics, from a set of text documents where topics are typically extracted as lists of weighted terms. A commonly employed implementation is latent Dirichlet allocation (LDA) [30], which is, for instance, used in the Europe Media Monitor (EMM) news aggregator [26].

Related articles can also be grouped with the help of *document clustering* methods, such as affinity propagation [91] or hierarchical agglomerative clustering (HAC) [226]. HAC, for example, computes pairwise document similarity on text features using measures such as the cosine distance on TF-IDF vectors [308] or word embeddings [197]. This way, HAC creates a hierarchy of the most similar documents and document groups [222]. HAC has been used successfully in several research projects [232, 276]. Other methods to group related articles exploit news-specific characteristics, such as the *five journalistic W questions* (5Ws). The 5Ws describe the main event of a news article, i.e., who did what, when, where, and why. A few works additionally extract the how question [321], i.e., how something happened or was done (5W1H extraction or question answering). Journalists usually answer the 5W questions within the first few sentences of a news article [52]. Once phrases answering the 5W question are extracted, articles can be grouped by comparing their 5W phrases. We propose a method for 5W1H extraction in Chap. 4.

News aggregation[1] is one of the most popular approaches to enable users to get an overview of the large amounts of news that is published nowadays. Established news aggregators, such as Google News and Yahoo News, show related articles by different outlets reporting on the same event. Hence, the approach is feasible to reveal instances of bias by source selection, e.g., if one outlet does not report on an important event. News aggregators rely on methods from computer science, particularly methods from natural language processing (NLP). The analysis pipeline of most news aggregators aims to find the most important news topics and present them in a compressed form to users. The analysis pipeline typically involves the following tasks [84, 128]:

1. *Data gathering*, i.e., crawling articles from news websites.
2. *Article extraction* from website data, which is typically HTML or RSS.
3. *Grouping*, i.e., finding and grouping related articles reporting on the same topic or event.
4. *Summarization* of related articles.
5. *Visualization*, e.g., presenting the most important topics to users.

For the first two tasks, data gathering and article extraction, established and reliable methods exist, e.g., in the form of web crawling frameworks [246]. Articles can be extracted with naive approaches, such as website-specific wrappers [270], or more generic methods based on content heuristics [185]. Combined approaches perform both crawling and extracting and offer other functionality tailored to news analysis. In Sect. 3.5, we propose *news-please*, a web crawler and extractor for news articles, which extracts information from all news articles on a website, given only the root URL of the news outlet to be crawled.

---

[1] The paragraphs about news aggregation have been adapted partially from [129].

The objective of grouping is to identify topics and group articles on the same topic, e.g., using LDA or other topic modeling techniques, as described previously. Articles are then summarized using methods such as simple TF-IDF-based scores or complex approaches considering redundancy and order of appearance [294]. By performing the five tasks of the news aggregation pipeline in an automatized fashion, news aggregators can cope with the large amount of information produced by news outlets every day.

However, no established news aggregator reveals event selection bias of news outlets to their users. Incorporating this functionality for short-term or event-oriented analysis of event selection bias, news aggregators could show the publishers that did *not* publish an article on a selected event. For long-term or trend-oriented analysis, news aggregators could visualize a news outlet's coverage frequency of specific topics, e.g., to show whether the issues of a specific politician or party or an oil company's accident is either promoted or demoted.

In addition to traditional news aggregators, which show topics and related topics in a list, recent news aggregators use different analysis approaches and visualizations to promote differences in news coverage caused by biased event selection. *Matrix-based news aggregation* (MNA) is an approach we devised earlier that follows the analysis workflow of established news aggregators while organizing and visualizing articles into rows and columns of a two-dimensional matrix [128, 129]. The exemplary matrix depicted in Fig. 2.2 reveals what is primarily stated by media in one country (rows) about another country (columns). For instance, the cell of the publisher country Russia and the mentioned country Ukraine, denoted with RU-UA, contains all articles that have been published in Russia and mention Ukraine. Each cell shows the title of the most representative article, determined through a TF-IDF-based summarization score among all cell articles [128]. Users either select rows and columns from a list of given configurations for common use cases, e.g., to analyze only major Western countries, or define own rows and columns from which the matrix shall be generated.

To analyze event selection bias, users can use MNA to explore main topics in different countries as in Fig. 2.2 or span the matrix over publishers and topics in a country.

Research in the social sciences concerned with bias by event selection requires significant effort due to the time-consuming manual linking of events from news articles to a second "baseline" dataset. Many established studies use event data from a source that is considered "objective," for example, police reports (cf. [6, 231, 267]). However, the automated extraction of relevant information from such non-news sources requires the development and maintenance of specialized tools for each of the sources. Reasons for the increased extraction effort include the diversity or unavailability of such sources, e.g., police reports are structured differently in different countries or may not be published at all. Linking events from different sources in an automated fashion poses another challenge because of the different ways in which the same event may be described by each of the sources. This places a limit on the possible contributions of automated approaches for comparison across sources or articles.

Publisher Countries	Mentioned Countries			
	UA	RU	GB	DE
RU	Foreign Policy Adviser Says Russia Committed to Peace Process in East Ukraine	Ukraine Crisis, Sanctions Against Russia Not on G20 Agenda in Australia: Russian Sherpa	Cameron Says Britain Will Pay Only Half of $2.6 Bln EU Surcharge	Berlin wall: the symbol of Cold War as an art object
GB	Ukraine crisis: Kiev accuses Russia of military invasion after ‚tanks cross border'	Tank column crosses from Russia into Ukraine – Kiev military	Cameron has warned there wil be a „major problem" if Brussels insists on Britain paying its $2.6 bn	Fall of the Berlin Wall: ‚Our tears of frustration turned to those of joy'
DE	Kyiv calls Berlin amid Russian incursion reports	Kyiv: 32 tanks enter Ukraine from Russia	Britain allowed to halve EU budget bill	Germany's east still lags behind
US	Ukraine accuses Russia of sending in donzens of tanks	Ukraine accuses Russia of sending in donzens of tanks	Britain finds deal with EU over controversial bill	AP WAS THERE: The Berlin Wall crumbles

Fig. 2.2 News overview to enable comparative news analysis in matrix-based news aggregation. The color of each cell refers to its main topic. Source [135]

In our view, the automated analysis of events within news articles, however, is a very promising line of inquiry for computer science research. Sophisticated tools can already gather and extract relevant data from online news sources. Methods to link events in news articles are already available or are the subject of active research [26, 30, 164, 222, 232, 276, 308]. In Sect. 4.2, we propose a method that extracts phrases describing journalistic properties of an article's main event, i.e., who did what, when, where, why, and how. Of course, news articles must originate from a carefully selected set of news publishers, which represent not only mainstream media but also alternative and independent publishers, such as Wikinews.[2] Finally, revealing differences in the selection of top news stories between publishers, or even the mass media of different countries, has shown promising results [128] and could eventually be integrated into regular news consumption using news aggregators demonstrating the potential for computer science approaches to make a unique contribution to the study event selection.

---

[2] https://en.wikinews.org/wiki/.

2.3.2   Source Selection

Journalists must decide on the trustworthiness of information sources and the actuality of information for a selected event. While source selection is a necessary task to avoid information overload, it may lead to biased coverage, e.g., if journalists mainly consult sources supporting one perspective when writing the article. The choice of sources used by a journalist or an outlet as a whole can reveal patterns of media bias. However, journalistic writing standards do not require journalists to list sources [371], which make the identification of original sources difficult or even impossible. One can only find hints in an article, such as the use of quotes, references to studies, phrases such as "according to [name of other news outlet]" [116], or the dateline, which indicates whether and from which press agency the article was copy-edited. One can also analyze whether the content and the argumentation structure match those of an earlier article [68].

The effects of source selection bias are similar to the effects of commission and omission (Sect. 2.3.3), because using only sources supporting one side of the event when writing an article (source selection) is similar to omitting all information supporting the other side (omission). Because many studies in the social sciences are concerned with the *effects* of media bias, e.g., [24, 69, 72, 98, 100, 159, 166, 190, 194, 237, 259, 399], and the effects of these three bias forms are similar, bias by source selection and bias by commission and omission are often analyzed together.

Few analyses in the social sciences aim to find the selected sources to derive insights on the source selection bias of an article or an outlet. However, there are notable exceptions, for example, one study counts how often news outlets and politicians cite phrases originating in think tanks and other political organizations. The researchers had previously assigned the organizations to a political spectrum [117]. The frequencies of specific phrases used in articles, such as "We are initiating this boycott, because we believe that it is racist to fly the Confederate Flag on the state capitol" [117], which originated in the civil rights organization NAACP, are then aggregated to estimate the bias of news outlets. In another study of media content, Papacharissi and Oliveira annotate indications of source selection in news articles, such as whether an article refers to a study conducted by the government or independent scientists [274]. One of their key findings is that UK news outlets often referred to other news articles, whereas US news outlets did that less often but referred to governments, opinions, and analyses.

On *social media*, people can be subject to their *own* source selection bias, as discussed in Sect. 2.1. For instance, on Facebook, people tend to be friends with likewise-minded people, e.g., who share similar believes or political orientations [15]. People who use social media platforms as their primary news source are subject to selection bias not only by the operating company [82, 85] but also by their friends [15].

To our knowledge, there are currently no approaches in computer science that aim to specifically identify bias by source selection. One exception is NewsDeps, an exploratory approach for determining the content dependencies between news articles [139]. Our approach employs simple methods from plagiarism detection (PD) described afterward to identify which parts of a news article stem from previously published news articles.

However, several automated techniques are well suited to address this form of bias. *Plagiarism detection* (PD) is a field in computer science with the broad aim of identifying instances of unauthorized information reuse in documents. Methods from PD may be used to identify the potential *sources* of information for a given article beyond identifying actual "news plagiarism" (cf. [179]). While there are some approaches focused on detecting instances of plagiarism in news, e.g., using simple text-matching methods to find 1:1 duplicates [309], research on news plagiarism is not as active as research on academic plagiarism. This is most likely a consequence of the fact that authorized copy-editing is a fundamental component in the production of news. Another relevant field that we describe in this section is *semantic textual similarity* (STS), which measures the semantic equivalence of two (usually short) texts [5].

The vast majority of *plagiarism detection* techniques analyzes text [89, 235] and thus could also be adapted and subsequently applied to news texts. Current methods can reliably detect *copy and paste* plagiarism, the most common form of plagiarism [89, 405]. *Ranking* methods use, for instance, TF-IDF and other information retrieval techniques to estimate the relevance of other documents as plagiarism candidates [149]. *Fingerprinting* methods generate hashes of phrases or documents. Documents with similar hashes indicate plagiarism candidates [149, 324]. *Hybrid* approaches assess documents' similarity using diverse features [236].

Today's plagiarism detection methods already provide most of the functionality to identify the potential sources of news articles. Copy-edited articles are often shortened or slightly modified and, in some cases, are a 1:1 duplicate of a press agency release. These types of slight modifications, however, can be reliably detected with ranking or fingerprinting methods (cf. [235, 309]). Current methods only continue to struggle with heavily paraphrased texts [235], but research is extending also to other non-textual data types such as analyzing links [107], an approach that can be used for the analysis of online news texts as well. Another text-independent approach to plagiarism detection are *citation-based* plagiarism detection algorithms, which achieve good results by comparing patterns of citations between two scientific documents [105]. Due to their text independence, these algorithms also allow a cross-lingual detection of information reuse [105]. News articles typically do not contain citations, but the *patterns* of quotes, hyperlinks, or other named entities can also be used as a suitable marker to measure the semantic similarity of news articles (cf. [107, 117, 203]). Some articles also contain explicit referral phrases, such as "according to *The New York Times.*" The *dateline* of an article can also state whether and from where an article was copy-edited [140]. Text search and rule-based methods can be used to identify referral phrases and to extract the resources being referenced. In our view, future research should focus on identifying the span of information that was taken from the referred resource (see also Sect. 2.3.3).

Semantic textual similarity (STS) methods measure the semantic equivalence of two (usually short) texts [5]. STS methods use basic measures, such as n-gram overlap, WordNet node-to-node distance, and syntax features, e.g., compare whether the predicate is the same in two sentences [312]. More recent methods combine various techniques and use deep learning networks, achieving a Pearson correlation of their STS results to human coders of 0.78 [306]. Recently, these methods have also focused on *cross-lingual* STS [5] and use, for example, machine translation before employing regular mono-lingual STS methods [36]. Machine translation has proven useful also for other cross-lingual tasks, such as event analysis [368].

Graph analysis is concerned with the analysis of relations between nodes in a graph. The relation between news articles can be used to construct a dependency graph. Spitz and Gertz analyzed how information propagates in online news coverage using hyperlinks linking to other websites [333]. They identified four types of hyperlinks: *navigational* (menu structure to navigate the website), *advertisement*, *references* (links within the article pointing to semantically related sites), and *internal links* (further articles published by the same news outlet). They only used reference links to build a network, since the other link types contain too many unrelated sites (internal) or irrelevant information (advertisement and navigational). One finding by Spitz and Gertz is that networks of news articles can be analyzed with methods of citation network analysis. Another method extracts quotes attributed to individuals in news articles to follow how information propagates over time in a news landscape [203]. One finding is that quotes undergo variation over time but remain recognizable with automated methods [203].

In our view, computer science research could therefore provide promising solutions to long-standing technical problems in the systematic study of source selection by combining methods from PD and graph analysis. If two articles are strongly similar, the later published article will most likely contain reused information from the former published article. This is a typical case in news coverage, e.g., many news outlets copy-edit articles from press agencies or other major news outlets [358]. Using PD, such as fingerprinting and pattern-based analysis as previously described, to measure the likelihood of information reuse between all possible pairs of articles in a set of related articles implicitly constructs a directed dependency graph. The nodes represent single articles, the directed edges represent the flow of information reuse, and the weights of the edges represent the degree of information reuse. The graph can be analyzed with the help of methods from graph analysis, e.g., to estimate importance or slant of news outlets or to identify clusters of articles or outlets that frame an event in a similar manner (cf. [333]). For instance, if many news outlets reuse information from a specific news outlet, the higher we can rate its importance. The detection of semantic (near-)duplicates would also help lower the number of articles that researchers from the social sciences need to manually investigate to analyze other forms of media bias in content analyses.

In summary, the analysis of bias by source selection is challenging, since the sources of information are mostly not documented in news articles. Hence, both in the social sciences and in computer science research, only few studies have analyzed this form of bias. Notable exceptions are the studies discussed previously that analyzed quotes used by politicians originating from think tanks. Methods from computer science can in principle provide the required techniques for the (semi-)automated analysis of this form of bias and thus make a very valuable contribution. The methods, most importantly those from plagiarism detection research, could be (and partially already have been [309]) adapted and extended from academic plagiarism detection and other domains, where reliable methods already exist.

2.3.3   Commission and Omission of Information

Analyses of bias by commission and omission compare the information contained in a news article with those in other news articles or sources, such as police reports and other official reports. The "implementation" and effects of commission and omission overlap with those of source selection, i.e., when information supporting or opposing a perspective is either included or left out of an article. Analyses in the social sciences aim to determine which frames the information included in such articles support. For instance, frame analyses typically compare the frequencies of frame-attributing phrases in a set of news articles [98, 120]. More generally, content analysis compares which facts are presented in news articles and other sources [326]. In the following, we describe exemplary studies of each of the two forms.

A frame analysis by Gentzkow and Shapiro quantified phrases that may sway readers to one or the other side of a political issue [98]. For this analysis, the researchers first examined which phrases were used significantly more often by politicians of one party over another and vice versa. Afterward, they counted the occurrence of phrases in news outlets to estimate the outlet's bias toward one side of the political spectrum. The results of the study showed that news producers have economic motives to bias their coverage toward the ideological views of their readers. Similarly, another method, briefly mentioned in Sect. 2.3.2, counts how often US congressmen use the phrases coined by think tanks, which the researchers previously associated with political parties [117]. One finding is that Fox News coverage was significantly slanted toward the US Republican Party.

A content analysis conducted by Smith et al. [326] investigated whether the aims of protesters corresponded to the way in which news reported one demonstrations. One of their key hypotheses was that news outlets will tend to favor the positions of the government over the positions of protesters. In the analysis, Smith et al. extracted relevant textual information from news articles, transcripts of TV broadcasts, and police reports. They then asked analysts to annotate the data and could statistically confirm the previously mentioned hypothesis.

Bias by commission and omission has not specifically been addressed by automated approaches despite the existence of various methods that we consider beneficial for the analysis of both forms of bias in a (semi-)automated manner. Researchers from the social sciences are already using text search to find relevant documents and phrases within documents [336]. However, search terms need to be constructed manually, and the final analysis still requires a human interpretation of the text to answer coding tasks, such as "assess the spin of the coverage of the event" [326]. Another challenge is that content analyses comparing news articles with other sources require the development of scrapers and information extractors tailored specifically to these sources.[3] To our knowledge, there are no established or publicly available generic extractors for commonly used sources such as police reports.

An approach that partially addresses commission and omission of information is *aspect-level browsing* as implemented in the news aggregator *NewsCube* [276]. Park et al. [276] define an "aspect" as the semantic proposition of a news topic. The aspect-level browsing enables users to view different perspectives on political topics. The approach follows the news aggregation workflow described in Sect. 2.3.1, but with a novel grouping phase: NewsCube extracts aspects from each article using keywords and syntactic rules and weighs these aspects according to their position in the article (motivated by the *inverted pyramid* concept: the earlier the information appears in the article, the more important it is [52]). Afterward, NewsCube performs HAC to group related articles. The visualization is similar to the topic list shown in established news aggregators, but additionally shows different aspects of a selected topic. A user study found that users of NewsCube became aware of the different perspectives and subsequently read more articles containing perspective-attributing aspects. However, the approach cannot reliably assess the diversity of the aspects. NewsCube shows all aspects, even though many of them are similar, which decreases the efficiency of using the visualization to get an overview of the different perspectives in news coverage. Word and phrase embeddings might be used to recognize the similarity of aspects (cf. [197, 319]). The visualization also does not highlight which information is subject to commission and omission bias, i.e., what information is contained in one article and left out in another article.

Methods from plagiarism detection (see Sect. 2.3.2) open a promising research direction for the automated detection of commission and omission of information in news. More than 80% of related news articles add no new information and only reuse information contained in previously published articles [358]. Comparing the presented facts of one article with the facts presented in previously published articles would help identify commission and omission of information. Methods from PD can detect and visualize which segments of a text may have been taken from other texts [105]. The relatedness of bias by source selection and bias by commission and omission suggests that an analysis workflow may ideally integrate methods from PD to address both issues (also see Sect. 2.3.2).

Centering resonance analysis (CRA) aims to find how influential terms are within a text by constructing a graph with each node representing a term that is contained in the noun phrases (NP) of a given text [60]. Two nodes are connected if their terms are in the same NP or boundary terms of two adjacent NPs. The idea of the approach is that the more edges a node has, the more influential its term is to the text's meaning. To compare two documents, methods from

---

[3] In Sect. 3.5, we propose a system for crawling and extracting news articles.

graph analysis can be used to analyze both CRA graphs (Sect. 2.3.2 gives a brief introduction to methods from graph analysis). Researchers from the social sciences have successfully employed CRA to extract influential words from articles and then manually compare the information contained in the articles [274]. Recent advancements toward computational extraction and representation of the "meaning" of words and phrases, especially word embeddings [197], may serve as another way to (semi-)automatically compare the contents of multiple news articles.

To conclude, studies in the social sciences researching bias by commission and omission have always compared the analyzed articles with other news articles and/or non-media sources, such as police reports. No approaches from computer science research specifically aim to identify this bias form. However, automated methods, specifically PD, CRA, graph analysis, and more recent also word embeddings, are promising candidates to address this form of bias opening new avenues for unique contributions of well-established computer science methodology in this area. CRA, for instance, has already been employed by researchers from the social sciences to compare the information contained in two articles.

2.3.4 Word Choice and Labeling

When referring to a semantic concept, such as an entity, a geographic position, or an activity, authors can *label* the concept and *choose from various words* to refer to it (cf. [86]). Instances of bias by labeling and word choice frame the referred concept differently, e.g., simply positively or negatively, or they highlight a specific perspective, e.g., economical or cultural (see Sect. 2.2.2 for a background on framing). Examples include "immigrant" or "economic migrant" and "Robert and John got in a fight" and "Robert attacked John." The effects of this form of bias range from concept level, e.g., a specific politician is shown to be incompetent, to article level, e.g., the overall tone of the article features emotional or factual words [263, 274].

Content analyses and framing analyses are used in the social sciences to identify bias by labeling and word choice within news articles. Similar to the approaches discussed in previous sections, the manual coding task is once again time-consuming, since annotating news articles requires careful human interpretation. The analyses are typically either *topic-oriented* or *person-oriented*. For instance, Papacharissi and Oliveira used CRA to extract influential words (see Sect. 2.3.3). They investigated labeling and word choice in the coverage of different news outlets on topics related to terrorism [274]. They found that *The New York Times* used a more dramatic tone, e.g., news articles dehumanized terrorists by not ascribing any motive to terrorist attacks or usage of metaphors, such as "David and Goliath" [274]. *The Washington Post* used a less dramatic tone, and both the *Financial Times* and *The Guardian* focused their news articles on factual reporting. Another study analyzed whether articles portrayed Bill Clinton, the US president at that time, positively, neutrally, or negatively [263].

The automated analysis of labeling and word choice in news texts is challenging due to limitations of current NLP methods [128], which cannot reliably interpret the frame induced by labeling and word choice, due to the frame's dependency on the context of the words in the text [266]. Few automated methods from computer science have been proposed to identify bias induced by labeling and word choice. Grefenstette et al. devised a system that investigates the frequency of affective words close to words defined by the user, for example, names of politicians [116]. They find that the automatically derived polarity scores of named entities are in line with the publicly assumed slant of analyzed news outlets, e.g., George Bush, the Republican US president at that time, was mentioned more positively in the conservative *The Washington Times* compared to other news outlets.

The most closely related field is *sentiment analysis*, which aims to extract an author's attitude toward a semantic concept mentioned in the text [272]. Current sentiment analysis methods reliably extract the unambiguously stated sentiment [272]. For example, those methods reliably identify whether customers used "positive," such as "good" and "durable," or "negative" words, such as "poor quality," to review a product [272]. However, the highly context-dependent, hence more ambiguous sentiment in news coverage described previously in this section remains challenging to detect reliably [266]. Recently, researchers proposed approaches using *affect analysis*, e.g., using more dimensions than polarity in sentiment analysis to extract and represent emotions induced by a text, and *crowdsourcing*, e.g., systems that ask users to rate and annotate phrases that induce bias by labeling and word choice [277]. We describe these fields in the following paragraphs.

While sentiment analysis presents one promising technique to be used for automating the identification of bias by word choice and labeling, the performance of current sentiment classification on news texts is poor (cf. [167, 266]) or even "useless" [335]. Two reasons why sentiment analysis performs poorly on news texts [266] are (1) the *lack of large-scale gold standard datasets* and (2) the *high context dependency* or implicitness of sentiment-inducing phrases. Large annotated datasets are required to train current sentiment classifiers [400]. More traditional classifiers use manually [153] or semi-automatically [13, 110, 335] created dictionaries of positive and negative words to score a sentence's sentiment. However, to our knowledge, no sentiment dictionary exists that is specifically designed for news texts, and generic dictionaries tend to perform poorly on such texts (cf. [16, 167, 266]). Godbole, Srinivasaiah, and Skiena [110] used WordNet to automatically expand a small, manually created seed dictionary to a larger dictionary. They used the semantic relations of WordNet to expand upon the manually added words to closely related words. An evaluation showed that the resulting dictionary had similar quality in sentiment analysis as solely manually created dictionaries. However, the performance of entity-related sentiment classification using the dictionary tested on news websites and blogs is missing a comparison against a ground truth, such as an annotated news dataset. Most importantly, dictionary-based approaches are not sufficient for news texts, since the sentiment of a phrase depends on its context, for example, in economics, a "low market price" may be good for consumers but bad for producers.

To avoid the difficulties of interpreting news texts, researchers have proposed approaches to perform sentiment analysis specifically on quotes [16] or on the comments of readers [278]. The motivation for analyzing only the sentiment contained in quotes or comments is that phrases stated by someone are far more likely to contain an explicit statement of sentiment or opinion-conveying words. While the analysis of quotes achieved poor results [16], readers' comments appeared to contain more explicitly stated opinions, and regular sentiment analysis methods perform better: a classifier that used the extracted sentiments from the readers' comments achieved a precision of 0.8 [278].

Overall, the performance of sentiment analysis on news texts is still rather poor. This is attributable to the fact that, thus far, not much research has focused on improving sentiment analysis when compared to the large number of publications targeting the prime use case of sentiment analysis: product reviews. Currently, no public annotated news dataset for sentiment analysis exists, which is a crucial requirement for driving forward successful, collaborative research on this topic.

A final challenge when applying sentiment analysis to news articles is that the one-dimensional positive-negative scale used by all mature sentiment analysis methods may fall short of representing the complexity of news articles. Some researchers suggested to investigate *emotions* or *affects*, e.g., induced by headlines [341] or in entire news articles [116], whereas investigating the full text seems to yield better results. *Affect analysis* aims to find the emotions that a text induces on the contained concepts, e.g., entities or activities, by comparing relevant words from the text, e.g., nearby the investigated concept, with affect dictionaries [344]. Bhowmick [28] devised an approach that automatically estimates which emotions a news text induces on its readers using features such as tokens, polarity, and semantic representation of tokens. An ML-based approach by Mishne classifies blog posts into emotion classes using features such as n-grams and semantic orientation to determine the mood of the author when writing the text [243].

Semantics derived using word embeddings may be used to determine whether words in an article contain a slant, since the most common word embedding models contain biases, particularly gender bias and racial discrimination [32, 42]. Bolukbasi describe a method to debias word embeddings [156]; the dimensions that were removed or changed by this process contain potentially biased words; hence, they may also be used to find biased words in news texts.

Besides fully automated approaches to identify bias by labeling and word choice, semi-automated approaches incorporate users' feedback. For instance, NewsCube 2.0 employs *crowdsourcing* to estimate the bias of articles reporting on a topic. The system allows users to collaboratively annotate bias by labeling and word choice in news articles [277]. Afterward, NewsCube 2.0 presents contrastive perspectives on the topic to users. In their user study, Park et al. [277] find that the NewsCube 2.0 supports participants to collectively organize news articles according to their slant of bias. Section 2.3.8 describes AllSides, a news aggregator that employs crowdsourcing, though not to identify bias by labeling and word choice but to identify spin bias, i.e., the overall slant of an article.

The forms of bias by labeling and word choice have been studied extensively in the social sciences using frame analyses and content analyses. However, to date, not much research on both forms has been conducted in computer science. Yet, the previously presented techniques from computer science, such as sentiment analysis and affect analysis, are already capable of achieving reliable results in other domains. Besides, crowdsourcing has already successfully been used to identify instances of such bias.

2.3.5   Placement and Size Allocation

The placement and size allocation of a story indicates the value a news outlet assigns to that story [14, 64]. Long-term analyses reveal patterns of bias, e.g., the favoring of specific topics or avoidance of others. Furthermore, the findings of such an analysis should be combined with frame analysis to give comprehensive insights on the bias of a news outlet, e.g., a news outlet might report disproportionately much on one topic, but otherwise, its articles are well-balanced and objective [75].

The first manual studies on the placement and size of news articles in the social sciences were already conducted in the 1960s. Researchers measured the size and the number of columns of articles present in newspapers, or the broadcast length in minutes dedicated to a specific topic, to investigate if there were any differences in US presidential election coverage [337–340]. These early studies, and also a more recent study conducted in 2000 [34], found no significant differences in article size between the news outlets analyzed. Fewer studies have focused on the placement of an article, but found that article placement does not reveal patterns of bias for specific news outlets [339, 340]. Related factors that have also been considered are the size of headlines and pictures (see also Sect. 2.3.6 for more information on the analysis of pictures), which also showed no significant patterns of bias [339, 340].

Bias by article placement and size has more recently not been revisited, even though the rise of online news and social media may have introduced significant changes. Traditional printed news articles are a permanent medium, in the sense that once they were printed, their content could not (easily) be altered, especially not for all issues ever printed. However, online news websites are often updated. For example, if a news story is still developing, the news article may be updated every few minutes (cf. [59]). Such updates of news articles also include the placement and allotted size of previews of articles on the main page and on other navigational pages. To our knowledge, no study has yet systematically analyzed the changes in the size and position of online news articles over time.

Fully automated methods are able to measure placement and size allocation of news articles because both forms can be determined by volumetric measurements (see Sect. 2.2.4). Printed newspapers must be digitalized first, e.g., using optical character recognition (OCR) and document segmentation techniques [160, 248]. Measuring a digitalized or online article's placement and size is a trivial task. Due to the Internet's inherent structure of linked websites, online news even allows for a more advanced and fully automated measurements of news article importance, such as PageRank [269], which could also be applied within pages of the publishing news outlet. Most popular news datasets, such as RCV1 [205], are text-based and do not contain information on the size and placement of a news article. Future research, however, should especially take into consideration the fast pace in online news production as described previously.

While measuring size and placement automatically is a straightforward task in computer science, only few specialized systems currently exist that can measure these forms of news bias. Saez-Trumper, Castillo, and Lalmas [307] devised a system that measures the importance ascribed to a news story by an outlet by counting the total number of words of news articles reporting on the story. To measure the importance ascribed to the story by the outlet's readers, the system counts the number of tweets linking to these news articles. One finding is that both factors are slightly correlated. NewsCube's visualization is designed to provide equal size and avoid unfair placement of news articles to "not skew users' visual attention" [276]. Even though the authors ascribe this issue high importance in their visualization, they do not analyze placement and size in the underlying articles.

Research in the social sciences and in computer science benefit from the increasing accessibility of online news, which allows effective automated analysis of bias by taking into consideration article placement and size. Measuring placement and size of articles is a trivial and scalable task that can be performed on any number of articles without requiring high manual effort. However, most recent studies in the social sciences have not considered including bias by placement and size into their analysis. The same is true for systems in computer science that should similarly include the placement and size of articles as an additional dimension of media bias. With the conclusions that have been drawn based on the analysis of traditional, printed articles, still in need of verification for online media, computer science approaches can here make a truly unique contribution.

2.3.6   Picture Selection

Pictures contained in news articles can influence how readers perceive a reported topic [304]. In particular, readers who wish to get an overview of current events are likely to browse many articles and thus view only each article's headline and image. The effects of picture selection even go so far as to influence readers' voting preferences in elections [304]. Reporters or news agencies sometimes (purposefully) show pictures out of context [83], e.g., a popular picture in 2015 showed an aggressive refugee with an alleged ISIS flag fighting against police officers. It later turned out that the picture was taken in 2012, before the rise of ISIS, and that the flag was not related to ISIS [70]; hence, the media had falsely linked the refugee with the terrorist organization.

Researchers from the social sciences have analyzed pictures used in news articles for over 50 years [173], approximately as long as media bias itself has been studied.

Basic studies count the number of pictures and their size to measure the degree of importance ascribed by the news outlet to a particular topic (see also Sect. 2.3.5 for information on bias by size). In this section, we describe the techniques studies use to analyze the semantics of selected images. To our knowledge, all bias-related studies in the social sciences are concerned with political topics. Analyses of picture selection are either *person-oriented* or *topic-oriented*.

Person-oriented analyses ask analysts to rate the articles' pictures showing specific politicians. Typical rating dimensions are [169, 371]:

- *Expression*, e.g., smiling vs. frowning
- *Activity*, e.g., shaking hands vs. sitting
- *Interaction*, e.g., cheering crowd vs. alone
- *Background*, e.g., the country's flags vs. not identifiable
- *Camera angle*, e.g., eye-level shots vs. shots from above
- *Body posture*, e.g., upright vs. bowed torso

Findings are mixed, e.g., a study from 1998 found no significant differences in the selected pictures between the news outlets analyzed, e.g., whether selected pictures of a specific news outlets were in favor of a specific politician [371]. Another study from 1988 found that *The Washington Post* did not contain significant picture selection bias but that *The Washington Times* selected images that were more likely favorable toward Republicans [169]. A study of German TV broadcasts in 1976 found that one candidate for German chancellorship, Helmut Schmidt, was significantly more often shown in favorable shots including better camera angles and reactions of citizens than the other main candidate, Helmut Kohl [171].

Topic-oriented analyses do not investigate bias toward persons but toward certain topics. For instance, a recent study on Belgian news coverage analyzed the presence of two frames [369]: asylum seekers in Belgium are (1) victims that need protection or (2) intruders that disturb Belgian culture and society. Articles supporting the first frame typically chose pictures depicting refugee families with young children in distress or expressing fear. Articles supporting the second frame chose pictures depicting large groups of mostly male, asylum seekers. The study found that the victim frame was predominantly adopted in Belgian news coverage and particularly in the French-speaking part of Belgium. The study also revealed a temporal pattern: during Christmas time, the victim frame was even more predominant.

To our knowledge, there are currently no systems or approaches from computer science that analyze media bias through image selection. However, methods in *computer vision* can measure many of the previously described dimensions. This is especially true since the recent rise of deep learning, where current methods achieve unprecedented classification performance [370]. Automated methods can identify faces in images, recognize emotions, categorize objects shown in pictures, and even generate captions for a picture. Research has advanced so far in these applications that several companies, such as Facebook, Microsoft, and Google, are using such automated methods in production, e.g., in autonomous cars, or are offering them as a paid service.

In the broad context of bias through image selection, Segalin et al. [317] trained a convolutional neural network (CNN) on the Psycho-Flickr dataset to estimate the personality traits of the pictures' authors. To evaluate the classification performance of the system, they compared the CNN's classifications with self-assessments by picture authors and also with attributed assessments by participants of a study. The results of their evaluation suggest that CNNs are suitable to derive such characteristics that are not even visible in the analyzed pictures.

Picture selection is an important factor in the perception of news. Basic research from psychology has shown that image selection can slant coverage toward one direction, although studies in the social sciences on bias by selection in the past concluded that there were no significant differences in picture selection. Advances in image processing research and the increasing accessibility of online news provide completely new avenues to study potential effects of picture selection. Computer science approaches can here primarily contribute by enabling the automated analysis of images on much bigger scale allowing us to reopen important questions on the effect of picture selection in news coverage and beyond.

2.3.7 Picture Explanation

Captions below images and referrals to the images in the main text provide images with the needed textual context. Images and their captions should be analyzed jointly because text can change a picture's meaning and vice versa [172, 173]. For instance, during Hurricane Katrina in 2005, two similar pictures published in US media showed survivors wading away with food from a grocery store. The only difference was that one picture showed a black man, who "looted" the store, while the other picture depicted a white couple, who "found" food in the store [328].

Researchers from the social sciences typically perform two types of analyses that are concerned with bias from image captions: jointly analyzing image and caption, or only analyzing the caption, ignoring the image. Only few studies analyze captions and images jointly. For instance, a comparison of images and captions from *The Washington Post* and *The Washington Times* found that the captions were not significantly biased [169]. A frame analysis on the refugee topic in Belgian news coverage also took into consideration image captions. However, the authors focused on the *overall* impression of the analyzed articles rather than examining any potential bias specifically present in the picture captions [369].

The vast majority of studies analyze captions without placing them in context with their pictures. Studies and techniques concerned with the text of a caption (but not the picture) are described in the previous sections, especially in the sections for bias by commission and omission (see Sect. 2.3.3) and labeling and word choice (see Sect. 2.3.4). We found that most studies in the social sciences either analyze image captions as a component of the main text or analyze images but disregard their captions entirely [339, 340, 371]. Likewise, relevant methods from computer science are effectively the same as those concerned with bias by commission and omission (see Sect. 2.3.3) and labeling and word choice (see Sect. 2.3.4). For the other type of studies, i.e., jointly analyzing images and captions, relevant methods are discussed in Sect. 2.3.6, i.e., computer vision to analyze the contents of pictures, and additionally in Sections 2.3.3 and 2.3.4, e.g., sentiment analysis to find biased words in captions.

To our knowledge, no study has examined picture referrals contained in the article's main text. This is most likely due to the infrequency of picture referrals.

The few analyses on captions suggest that bias by picture explanation is not very common. However, more fundamental studies show the impact of captions on the perception of images and note rather subtle differences in word choice. While many studies analyzed captions as part of the regular text, e.g., analyzing bias by labeling and word choice, research currently lacks specialized analyses that examines captions in conjunction with their images.

2.3.8   Spin: The Vagueness of Media Bias

Bias by spin is closely related to all other forms of media bias and is also the vaguest form. Spin is concerned with the context of presented information. Journalists create the spin of an article on all textual levels, e.g., by supporting a quote with an explanation (phrase level), by highlighting certain parts of the event (paragraph level), or even by concluding the article with a statement that frames all previously presented information differently (article level). The order in which facts are presented to the reader influences what is perceived (e.g., some readers might only read the headline and lead paragraph) and how readers rate the importance of reported information [52]. Not only the text of an article but all other elements, including pictures, captions, and the presentation of the information, contribute to an article's overall spin.

In the social sciences, the two primarily used methods to analyze the spin of articles are frame analysis and more generally content analysis. For instance, one finding in the terrorism analysis conducted by Papacharissi and Oliveira (see Sect. 2.3.2) was that *The New York Times* often personified the events in their articles, e.g., by focusing on persons involved in the event and the use of dramatic language [274].

Some practices in *journalism* can be seen as countermeasures to mitigate media bias. *Press reviews* summarize an event by referring to the main statements found in articles by other news outlets. This does not necessarily reveal media bias, because any perspective can be supported by source selection, e.g., only "reputable" outlets are used. However, typically press reviews broaden a reader's understanding of an event and might be a starting point for further research. Another practice that supports mitigation of media bias are opposing commentaries in newspapers, where two authors subjectively elaborate their perspective on the same topic. Readers will see both perspectives and can make their own decisions regarding the topic.

Social media has given rise to new collaborative approaches to media bias detection. Reddit[4] is a social news aggregator, where users post links or texts regarding current events or other topics and rate or comment on posts by other users. Through the comments on a post, a discussion can emerge that is often controversial and contains the various perspectives of commenters on the topic. Reddit also has a "media bias" thread[5] where contributors share examples of biased articles. Wikinews[6] is a collaborative news producer, where volunteers author and edit articles. Wikinews aims to provide "reliable, unbiased and relevant news [...] from a neutral point of view." However, two main issues are as follows: first, the mixed quality of the news items, because many authors may participate in producing them, and second, the low number of articles, i.e., only major events are covered in the English version and other languages have even fewer articles. Thus, Wikinews currently cannot be used as a primary, fully reliable news source. Some approaches employ crowdsourcing to visualize different opinions or statements on politicians or news topics, for example, the German news outlet Spiegel Online frequently asks readers to define their position regarding two pairs of contrary statements that span a two-dimensional map [331]. Below the map, the news outlet lists excerpts from other outlets that support or contradict the map's statements.

The automated analysis of spin bias using methods from computer science is maybe the most challenging of all forms because its manifestation is the vaguest among the forms of bias discussed. Spin refers to the overall perception of an article. Bias by spin is not, however, just the sum of all other forms but includes other factors, such as the order of information presented in a news article, the article's tone, and emphasis on certain facts. Methods we describe in the following are partially also relevant for other forms of bias. For instance, the measurement of an article's degree of personification in the terrorism in news coverage study [274] is supported by the computation of CRA [52]. What is not automated is the annotation of entities and their association with an issue. Named entity extraction [255, 391] could be used to partially address these previously manually performed tasks.

Other approaches analyze news readers' input, such as readers' comments, to identify differences in news coverage. The rationale of these approaches is that readers' input contains explicitly stated opinions and sentiment on certain topic, which are usually missing from the news article itself. Explicitly stated opinion can reliably be extracted with the help of NLP methods, such as sentiment analysis. For instance, one method analyzes readers' comments to categorize related articles [1]. The method measures the similarity of two articles by comparing their reader comments, thereby focusing in each comment on the mentioned entities, the expressed sentiment, and country of the comment's author. Another method counts and analyzes Twitter followers of news outlets to estimate the political orientation of the audience of the news outlet [111]. A seed group of Twitter accounts is

---

[4] https://www.reddit.com/.

[5] https://www.reddit.com/r/MediaBias/.

[6] https://en.wikinews.org/wiki/.

manually rated according to their political orientation, e.g., conservative or liberal. This group is automatically expanded using those accounts' followers. The method then estimates the political orientation of a news outlet's audience by averaging the political orientation of the outlet's followers in the expanded group of categorized accounts (cf. [98, 117, 220]).

The news aggregator *AllSides* [8] shows users the most contrastive articles on a topic, e.g., left and right leaning on a political spectrum. The system asks users to rate the spin of news outlets, e.g., after reading articles published by these outlets. To estimate the spin of an outlet, AllSides uses the feedback of users and expert knowledge provided by their staff. *NewsCube 2.0* lets (expert) users collaboratively define and rate frames in related articles [277]. The frames are in turn presented to other users, e.g., a contrast view shows the most contrasting frames of one event. Users can then incrementally improve the quality of coding by refining existing frames.

Another method for news spin identification categorizes news articles on contentious news topics into two (opposing) groups by analyzing quotes and nearby entities [275]. The rationale of the approach is that articles portraying a similar perspective on a topic have more common quotes, which may support the given perspective, than articles that have different perspectives. The method extracts weighted triples representing who criticizes whom, where the weight depends on the importance of the triple, e.g., estimated by the position within the article (the earlier, the more important). The method measures the similarity of two articles by comparing their triples.

Other methods analyze frequencies and co-occurrences of terms to find frames in related articles and assign each article to one of the frames. For instance, one method clusters articles by measuring the similarity of two documents using the co-occurrences of the two documents' most frequent terms [241]. The results of this rather simple method are then used for a manually conducted frame analysis. *Hiérarchie* uses recursive topic modeling to find topics and subtopics in tweets posted by users on a specific issue [325]. A radial treemap visualizes the extracted topics and subtopics. In the presented case study, users find and explore different theories on the disappearance of flight MH-370 discussed in tweets.

Lastly, manually annotated information related to media bias, e.g., the overall spin of articles rated by users of AllSides or articles annotated by social scientists during frame analysis, can in our view serve as a basis when creating training datasets for *machine learning*. Other data that exploits the *wisdom of the crowd* might be incorporated as well, e.g., analyzing the Reddit media bias thread. However, one should carefully review the information for its characteristics and inherent biases, especially if crowdsourced.

In our view, the existence of the very concept of spin bias allows drawing two conclusions. First, media bias is a complex model of skewed news coverage with overlapping and partially contradicting definitions. While many instances of media bias fit into one of the other more precisely defined forms of media defined in the news production and consumption process (see Sect. 2.2.3), some instances of bias do not. Likewise, such instances of bias may fit into other models from the social sciences that are concerned with differences in news coverage, such as the bias forms of coverage, gatekeeping, and statement (Sect. 2.2.3 briefly discusses other models of media bias), while other instances would not fit into such models. Second, we found that most of the approaches from computer science for identifying, or suitable for identifying, spin bias omit the research that has been conducted in the social sciences. Computer science approaches currently still address media bias as vaguely defined differences in news coverage and therefore stand to profit from prior research in the social sciences. In turn, there are few scalable approaches to the analysis of media biases in the social sciences significantly hampering progress in the field. We therefore see a strong prospect for collaborative research on automated approaches to the analysis of media bias across both disciplines.

2.3.9 Summary

Most automated approaches focus on analyzing vaguely defined "biases." These biases can be technically significant but may often not represent meaningful slants of the news. In contrast, in social science research, media bias emerges from observing systematic tendencies of specific bias forms or means. For example, the news production process that we use in our literature review defines nine bias forms.

One reason for the previously mentioned lack of conclusive or meaningful results is that almost no automated approach aims to specifically find such individual bias forms. At the same time, however, we found that suitable automated techniques are available to aid in the analysis of the individual bias forms.

2.4 Reliability, Generalizability, and Evaluation

This section discusses how automated approaches for analyzing media bias should be evaluated. Therefore, we first describe how social scientists measure the reliability and generalizability of studies on media bias.

The reliability and generalizability of manual annotation in the social sciences provide the benchmark for any automated approach. Best practices in social science research can involve both the careful development and iterative refinement of underlying codebooks and the formal validation of inter-coder reliability. For example, as discussed in Sect. 2.2.4, a smaller, careful inductive manual annotation aids in constructing the codebook. The main deductive analysis is then performed by a larger pool of coders where the results of individual coders and their agreement on the assignment of codes can be systematically compared. Standard measures for inter-coder reliability, e.g., the widely used Krippendorff's alpha [144], provide estimates for the reliability and robustness of the coding. Whether coding rules, and with these the quality of annotations, can be generalized beyond a specific case is usually not routinely analyzed because, by virtue of the significant effort required for manual annotation, the scope of such studies is usually limited to a specific question or context. Note, however, that the usual setup of a small deductive analysis, conducted on a subset of the data, implies that a codebook generated in this way can generalize to a larger corpus.

Computer science approaches for the automated analysis of media bias stand to profit a lot from a broad adoption of their methods by researchers across a wider set of disciplines. The impact and usefulness of automatized approaches for substantial cross-disciplinary analyses, however, hinge critically on two central questions. First, compared to manual methodologies, how reliable are automated approaches? Specifically, broad adoption of automated approaches in social sciences applications is only likely if the automated approaches identify at least close to the same instances of bias as manual annotations would.

Depending on which kind of more or less subtle form of bias is analyzed, the results gained through manual annotation might represent a more or less difficult benchmark to beat. Especially in complex cases, manual annotation of individual items may systematically perform better in capturing subtle instances relevant to the analysis question than automated approaches. Note that, for example, currently no public annotated news dataset for sentiment analysis exists (see Sect. 3.4). The situation is similar for most of the applications reviewed in this chapter, i.e., there is currently a dearth of standard benchmark datasets. Meaningful validation would thus require as a first step the careful (and time-intensive) development of such datasets across a range of relevant contexts.

One way to counter the present lack of evaluation datasets is to not solely rely on manual content analysis for annotation. For simple annotation tasks, such as rating the subjective slant of a news picture, crowdsourcing can be a suitable alternative to content analysis. This procedure requires less effort than conducting a full content analysis, including creation of a codebook and refining it until the ICR is sufficiently high (cf. [152]). One can also use other available data. For example, Recasens, Danescu-Niculescu-Mizil, and Jurafsky [297] use bias-related revisions from the Wikipedia edit history to retrieve presumably biased single-word phrases. The political slant classification of news articles and outlets crowdsourced by users on web services such as AllSides (see Sect. 2.3.8) may serve as another comparison baseline. As stated in Sect. 2.3.8, before employing crowdsourced information, one should carefully review its characteristics and quality.

Another way to evaluate the performance of bias identification methods is to manually analyze the automatically extracted instances of media bias, e.g., through crowdsourcing or (typically fewer) specialized coders. However, evaluating the results of an automated approach this way decreases the comparability between approaches, since these have to be evaluated in the same way manually again. Generating annotated benchmark datasets on the other hand requires greater initial effort, but the results can then be used multiple times to evaluate and compare multiple approaches.[7]

The second central question is how well-automated approaches generalize to the study of similar forms of bias in different contexts than those contexts for which they were initially developed. This question pertains to the external validity of developed approaches, i.e., is their performance dependent on a specific empirical or topical context? Out-of-sample performance could be tested against benchmark datasets not used for initial evaluation; however, as emphasized before, such datasets must still be developed. Hence, systematically testing the performance of approaches across many contexts is likely also infeasible for the near future simply because the cost of generating benchmark datasets is too high. Ultimately, it would be best practice for benchmark studies to establish more generally whether or not specific characteristics of news are related to the performance of the automated approaches developed.

2.5 Key Findings

News coverage strongly influences public opinion. While slanted news coverage is not harmful per se, systematically biased news coverage can negatively impact the public. Recent trends, such as social bots that automatically write news posts or the centralization of media outlet ownership, have the potential to further amplify the negative effects of biased news coverage. News consumers should be able to view different perspectives of the same news topic [252]. Unrestricted access to unbiased information is crucial for citizens to form their own views and make informed decisions [135, 250], e.g., during elections. Since media bias has been, and continues to be, structurally inherent in news coverage [146, 147, 276], the detection and analysis of media bias is a topic of high societal and policy relevance.

Researchers from the social sciences have studied media bias over the past decades, resulting in a comprehensive set of methodologies, such as content analysis and frame analysis, as well as models to describe media bias. One of these models, the *news production process*, describes how journalists turn events into news articles. The process defines nine forms of media bias that can occur during the three phases of news production: In the first phase, "gathering of information," the bias forms are (1) event selection, (2) source selection, and (3) commission and omission of information. In the second phase, "writing," the bias forms are (4) labeling and word choice. In the third phase, "editing," the bias forms are (5) story placement, (6) size allocation, (7) picture selection, and (8) picture explanation. Lastly, bias by (9) spin is a form of media bias that represents the overall bias of a news article

---

[7] The SemEval series [5] are a representative example from computer science where with high initial effort comprehensive evaluation datasets are created, allowing a quantitative comparison of the performance of multiple approaches afterward.

and essentially combines the other forms of bias, including minor forms not defined specifically by the news production and consumption process.

For each of the forms of media bias, we discussed exemplary approaches being applied in the social sciences and described the automated methods from computer science that have been used, or could best be used, to address the particular form of bias. We summarize the findings of our review of the current status quo as follows:

F1 Only few approaches in computer science address the analysis of media bias. The majority of these approaches analyze media bias from the perspective of regular news consumers and neglect both the approaches and models that have already been developed in the social sciences. In many cases, the underlying models of media bias are too simplistic, and their results when compared to models and results of research in the social sciences do not provide additional insights.

F2 The majority of content analyses in the social sciences do not employ state-of-the-art methods for automated text analysis. As a result, the manual content analysis approaches conducted by social scientists require exacting and very time-consuming effort, as well as significant expertise and experience. This severely limits the scope of what social scientists can study and has significantly hampered progress in the field.

F3 Thus, there is, in our view, much potential for interdisciplinary research on media bias among computer scientists and social scientists. Automated approaches are available for each of the nine forms of media bias that we discussed. On the one hand, methodologies and models of media bias in the social sciences can help to make automated approaches more effective. Likewise, the development of automated methods to identify instances of specific forms of media bias can help make content analysis in the social sciences more efficient by automating more tasks.

Media bias analysis is a rather young research topic within computer science, particularly when compared with the social sciences, where the first studies on media bias were published more than 70 years ago [172, 377]. Our first finding (F1) is that most of the reviewed computer science approaches treat media bias vaguely and view it only as "differences of [news] coverage" [278], "diverse opinions" [251], or "topic diversity" [252]. The majority of the current approaches neglect the state of the art developed in the social sciences. They do not make use of models describing different forms of media bias or how biased news coverage emerges in the news production and consumption process [14, 276] (Sect. 2.2.3). Also, approaches in computer science do not employ methods to analyze the specific forms of bias, such as content analysis [64] and frame analysis [368] (Sect. 2.2.4). Consequently, many approaches in computer science are limited in their capability for identifying instances of media bias. For instance, matrix-based news aggregation (MNA) organizes articles and topics in a matrix to facilitate showing differences in international news topics, but the approach can neither determine whether there are actual differences, nor can MNA enforce finding differences [129]. Likewise, *Hiérarchie* finds subtopics in news posts that may or may not refer to differences caused by media bias [325]. To overcome the limitations in identifying bias, some approaches, such as *NewsCube 2.0* [277] and *AllSides* (Sect. 2.3.8), outsource the task of identifying media bias to users, e.g., by asking users to manually rate the slant of news articles.

Content analysis and frame analysis both require significant manual effort and expertise (F2). Especially time-intensive are the tasks of systematic screening and subsequent annotation of texts. Such tasks can currently only be performed by human coders [64, 368]. Currently, in our view, the execution of these tasks cannot be improved significantly by employing automated text analysis methods due to the lack of mature methods capable of identifying specific instances of media bias, which follows from F1. This limitation, however, may be revised once interdisciplinary research has resulted in more advanced automated methods. Other tasks, such as data gathering or searching for relevant documents and phrases, are already supported by basic (semi-)automated methods and tools, such as content analysis software [215]. However, clearly the full potential of the state of the art in computer science is not yet being exploited. The employed techniques, e.g., keyword-based text matching to find relevant documents [336] or frequency-based extraction of representative terms to find patterns [215], are rather simple compared to state-of-the-art methods for text analysis. Few of the reviewed tools used by researchers in the social sciences employ methods proven effective in natural language processing, such as resolution of coreferences or synonyms or finding related article using an event-based search approach.

In our view, combining the expertise of the social sciences and computer science results in valuable opportunities for interdisciplinary research (F3). Reliable models of media bias and manual approaches for the detection of media bias can be combined with methods for automated data analysis, in particular, with text analysis and natural language processing approaches. NewsCube [276], for instance, extracts so-called aspects from news articles, which refer to the frames defined by social scientists [159]. Users of NewsCube became more aware of the different perspectives contained in news coverage on specific topics, than users of Google News. In this chapter, we showed that promising automated methods from computer science are available for all forms of media bias as defined by the news production and consumption process (see Sect. 2.3). For instance, studies concerned with bias by source selection or the commission and omission of information investigate how information is reused in news coverage [98, 117, 120]. Similarly to these studies, methods from plagiarism detection aim to identify instances of information reuse in a set of documents, and these methods yield reliable results for plagiarism with sufficient textual similarity [89, 179]. Finally, recent advancements in text analysis, particularly word embeddings [197] and deep learning [198], open a promising area of research on media bias. Thus far, few studies use word embeddings and deep learning to analyze media bias in news coverage. However, the techniques have proven very successful in various related problems (cf. [5, 191, 306, 311]), which lets us anticipate that the majority of the textual bias forms could be addressed effectively with such approaches.

We believe that interdisciplinary research on media bias can result in three main benefits. First, automated approaches for analyzing media bias will become more effective and more broadly applicable, since they build on the substantial, theoretical expertise that already exists in the social sciences. Second, content analyses in the social sciences will become more efficient, since more tasks can be automated or supported by automated methods from computer science. Finally, we argue that news consumers will benefit from improved automated methods for identifying media bias, since the methods can be used by news aggregators to detect and visualize the occurrence of potential media bias in real time.

2.6   Practical View on the Research Gap: A Real-World Example

This section practically demonstrates the implications of the literature review's finding using a real-world example of news coverage and consumption.

Objective Suppose you are reading the news. When viewing the coverage on an event, e.g., in your favorite news aggregator, or a single article reporting on the event, e.g., on the website of your favorite news outlet, you are wondering whether there might be other perspectives on the event. Which information are you missing since it is not mentioned in the articles you viewed or read? Mapping these questions to the terminology introduced earlier, the objective in this scenario is to efficiently and effectively get an overview of all the major perspectives present in the media. Efficiency is vital since newsreaders typically have only limited time for informing themselves on current events. While this example entails only one event, newsreaders are interested in multiple events, limiting the time available for a single event further. Effectiveness refers to understanding distinct and meaningful perspectives that help determine whether one already has a comprehensive overview of the coverage or if and which articles may offer alternative interpretations or additional information.

Setting Table 2.2 shows headlines of news articles reporting on the Republican Party debate during the US presidential primaries in New Hampshire hosted by ABC News on February 6, 2016. We selected the articles using the following criteria: they had to primarily report on the event and be published by a popular online US news outlet[8] on the day of the event or the day after. This way, we retrieved more than 30 articles. Afterward, we conducted an inductive frame analysis (Sect. 2.2.4.2) to get a comprehensive overview of the content and perspectives present in the event coverage. For the sake of simplicity in this example, we selected eight articles that represented all major perspectives with only minor differences between the articles.

---

[8] An outlet was defined as being "popular" if it was contained in the list of "top outlets" shown on https://www.allsides.com/media-bias/media-bias-ratings.

Table 2.2 Articles' headlines on an event of the 2016 Republican Party presidential primaries. Column "Pol." refers to the outlet's political orientation as stated on AllSides [8]. For a list of the articles' URLs, please refer to Table A.1 in Sect. A.1

ID	Outlet	Pol.	Headline
0	CNN	L	GOP takes aim at Marco Rubio during debate
1	PoliticusUSA	L	Here Are The Winners And Losers From ABC's Republican Presidential Debate
2	USA Today	C	Eighth GOP debate: Highlights from New Hampshire
3	Chicago Tribune	C	GOP debate: Trump calls for a lot worse than waterboarding against terrorists
4	CNBC	C	Rubio falters in presidential debate, offering hope to rivals
5	Fox News	R	Top tier takes heat: Rubio, others under fire at NH debate
6	Fox News	R	After strong debate, Christie, Bush resume attack on Rubio
7	Fox News	R	Marco Rubio is biggest loser. Trump and the governors all have a good night in NH

In daily news consumption, the eight articles could, for example, be the results of an online search for coverage on the event or be shown in a news aggregator or another news application. Note that our pre-selection of articles already gives an unrealistic improvement concerning the example's objective compared to regular news consumption because the article set is small and at the same time fully represents the coverage's substantial frames.

Interactive experiment

Look at the headlines in Table 2.2. The headlines are taken from news articles that report on a debate during the 2016 presidential primaries. Estimate how many major perspectives there are in the event coverage on the debate. Think of a perspective as a distinct viewpoint on the debate that is the most prominent viewpoint common to one or more articles.

Next, decide for each article which perspective it has on the event.

You can try to increase the "accuracy" of your results by looking at further information, such as the articles' outlets, their political orientation (Table 2.2), or the articles' full text (Appendix A.1). Please write down your results for each article and compare them with those presented in the following.

Manual Frame Analysis The previously mentioned frame analysis yielded three frames,[9] which are shown in the last column ("Frame") for each article ("ID") in Table 2.3. Frame F1 occurs in a single article (ID 2 with political orientation center), which is the only article that was updated consistently during the event to contain up-to-date information. In contrast to the other frames and articles, F1 consists primarily of quotes by the candidates, mostly about themselves. The frame thus portrays most candidates as they portrayed themselves in the debate, i.e., positively. There is not much commentary or assessment by journalists in this frame.

Common to much coverage on the event and thus also common to the two remaining frames is the prominence of three candidates. Chris Christie is portrayed as rather strong, and Marco Rubin as weak, being a target of verbal attacks by Christie and the other candidates. Also common to most articles reporting on the debate is that they prominently or often report on Donald Trump. At the time of the event, he generally received particular media interest, e.g., because he had boycotted the previous debate. As such, Trump is also frequently mentioned in the remaining articles of the set and serves as a distinguishing factor for the two remaining frames. Articles of frame F2 portray Trump rather negatively. Articles of F2 mention, for example, that Trump was "booed" by the audience (0, left), that Trump was accused "of taking advantage of an elderly woman" (3, center), and that "Trump was hit hard by Bush" (6, right). In contrast, articles of frame F3 portray Trump primarily positively, e.g., that "he seemed to do well enough to possibly win" (4, center), that "he was unwaveringly in charge" (7, right), that "Trump was measured and thoughtful" (7, right), and that "it is easy to see the Trump train getting on a roll" (1, left).

We use the results of the manual frame analysis as the ground truth since the technique represents one of the standards in social science research on media bias.

Means for Bias-Sensitive News Consumption In addition to frame analysis, we tested three means to identify the articles' perspectives. These means represent practices suitable for daily news consumption as well as automated techniques. Table 2.3 shows the perspectives assigned to individual articles by the approaches. The column "Headline" represents a means applied by many news consumers due to its high efficiency, i.e., determining the content of an article by its headline. Specifically, the column contains the author's results of the previous interactive experiment, where H1 represents a perspective[10] that portrays Rubin negatively. Using as much information as available in the headlines, we identified two sub-perspectives of H1 where additionally Christie and Bush are portrayed positively (H1a) and Trump is portrayed positively (H1b). H2 represents an "anti-Trump"

---

[9] Frame analyses are task-specific, and the resulting frames may depend on the data and analysis question at hand. Due to the articles' focus on persons involved in the debate, we centered our framing categories on these persons.

[10] We use the term "perspective" to highlight that this classification resulted from applying a practice or technique. In contrast to a frame, a perspective may, however, not fully or meaningfully represent an article's content and framing.

Table 2.3 Results of approaches to identify biases in the real-world example. The columns "Headline," "Political," "Clustering," and "Frame" show each article's central perspective on the event according to the headline's potential frame as identified by the author, the outlet's political orientation, an automated clustering technique on word embeddings, and inductive manual frame analysis. For each approach, the colors of its groups are chosen to maximize congruence with the framing groups of the inductive frame analysis. The higher the visual congruence of any column with the "Frame" column, the better

ID	Headline	Political	Clustering	Frame
0	H1	P1	C2	F2
1	?	P1	C2	F3
2	?	P2	C1	F1
3	H2	P2	C3	F2
4	H1	P2	C2	F3
5	H3	P3	C2	F2
6	H1(a)	P3	C2	F2
7	H1(b)	P3	C2	F3

perspective, and in perspective H3, all candidates and especially Rubio are portrayed negatively. Following the previous perspective categorization centered on persons, two headlines (articles 1 and 2) could not be assigned to a meaningful perspective.[11] When comparing these headline-implied perspectives with the frames in the right column that were deduced by carefully analyzing the articles' full content, the lack of an overall coherence across both directly indicates that the headlines do not allow for reliably estimation of an article's slant.

Using the political orientation of the articles' outlets to determine the articles' potential slant is another means [8] for bias identification (column "Political"). Employing the left-right dichotomy is fast and often also effective when analyzing political discourse and even more so in polarized media landscapes such as in the USA [395]. However, the lack of coherence between the perspectives implied by the outlets' political orientation and the frames shown in Table 2.3 highlights that this approach is superficial and its results are inconclusive. While employing the political orientation can increase the visibility of slants, they cannot reliably identify an article's slant. In the example, there are major differences even across articles that have the same perspective according to this means.

The clustering approach (column "Clustering"), albeit simply using affinity propagation [91] on word embeddings,[12] is the only approach to determine the previously mentioned difference of article 2, the only with frame F1, compared to all others. However, otherwise, the technique yields inconclusive results, e.g., a

---

[11] However, in another categorization scheme, the headlines could be interpreted as a perspective giving an overview of the event.

[12] The embeddings were derived using the largest model "en_core_web_lg" of the natural language processing toolkit spaCy (v3.0). Source: https://spacy.io/usage/v3.

large group of articles (C2) entailing articles from the entire political spectrum, and entails both remaining frames. The results of this simple approach are representative of automated approaches for bias identification, which analyze bias, for example, as vaguely defined "topic diversity" [252] or "differences in coverage" [278] as shown in the literature review. Other technical means may even amplify the newsreaders' own biases, e.g., Google News, Facebook, and other news aggregators or channels learn from users' preferences and show primarily those news items that are to the users' liking or interest.[13]

Summary Of course, the generalizability of this simple example is limited by various factors. For example, the inductive frame analysis was conducted only by one person, likely increasing the degree of subjectivity. In frame analyses, researchers in the social sciences typically rely on the annotations of multiple persons. At least during test phases, the annotations are compared and discussed to avoid subjectivity or achieve a known level of subjectivity that is coherent across the annotations (Sect. 2.2.4).

However, the example also highlights two key findings of our literature review. Whether they are automated or manual, current means are unreliable and suffer from superficial methodology and results or are reliable but cause high manual effort. There is no coherence across the perspectives determined by the three fast approaches compared to the results of the frame analysis. There is not even any coherence when comparing any pair of the fast methods.

If you participated in the interactive experiment, your findings might differ from those shown in Table 2.3, depending on which information you analyzed. Examining further information than the headlines alone may have yielded a more comprehensive understanding of the news coverage but came at an additional investment of time and effort. This effort is even increased in regular news consumption since newsreaders first have to research relevant articles of an event. Ultimately, critical assessment of the news takes too much time to be applied during regular news consumption. However, as automated approaches are unreliable, such manual practices currently present the only reliable means to analyze media bias.

It is this gap that the thesis at hand aims to address.

2.7   Summary of the Chapter

This chapter reviewed the issue of media bias and gave an interdisciplinary overview on the topic, particularly on methods and tools used to analyze media bias. The com-

---

[13] A typical example highlighting the filter bubble issue occurred when compiling the set of articles used in this example. Google News and Google Search presented the author with articles from only two political orientations, even when using the browser's privacy mode. This could only be overcome by using search engines that did not adapt search results to their users, such as DuckDuckGo.

parison of prior work in computer science, political science, and related disciplines revealed differences. Media bias has been studied extensively in the social sciences, whereas it is a relatively young research subject in computer science and other disciplines concerned with devising automated approaches. Consequently, while many automated methods offer effortless, scalable analysis, they yield inconclusive or less substantial results than methods used in the social sciences. Conversely, social science methods are practice-proven and effective but require much effort because researchers have to conduct them manually.

The chapter showed that the work conducted in either of the disciplines could benefit from incorporating knowledge and methods established in the other disciplines. Thus, while this thesis has a focus on computer science methodology, our general research principle is to make use of social science expertise where possible and feasible. Chapter 3 discusses how we can effectively address our research question in the context of the state of the art in computer science and the social sciences.

Chapter 3
Person-Oriented Framing Analysis

Abstract This chapter proposes person-oriented framing analysis (PFA), our approach to reveal biases. In a discussion of the solution space to tackle media bias, the chapter uses the findings of the literature review from Chap. 2 to narrow the research question from Chap. 1 down to a specific research objective. The PFA approach seeks to address this research objective. In contrast to prior work, PFA is designed to approximate analysis concepts and uses methodology established in the social sciences for bias analysis. As Chap. 6 shows, by identifying in-text means of specific bias forms, the approach detects meaningful frames in person-centric news coverage. Besides, this chapter introduces news-please, a crawler and extractor for online news articles that can be used in various use cases, such as prior to PFA.

3.1 Introduction

The previous chapters showed that media bias and its extreme form, fake news, are pressing issues that can influence how—and even if at all—societies can make decisions. Only rarely are news consumers aware of biases in the news. Revealing biases as such, for example by showing the frames present in coverage on the same topic, can help news consumers become aware of bias and make more informed decisions. This idea is at the heart of the research presented in this thesis.

Automatically Identifying Framing in News Articles to Reveal Bias

As the literature review in Chap. 2 showed, media bias is highly complex and—albeit in computer science often being analyzed as a single, rather broadly defined concept—consists of a broad spectrum of forms. Many of these forms are rather subtle and difficult to identify. During daily news consumption, sophisticated critical assessment of news coverage is nearly impossible if one is not trained to recognize such bias form *and* can afford to invest significant time, e.g., for researching facts and contrasting coverage. In sum, identifying framing or media bias using, for example, media literacy practices and social science frame analyses requires in-depth expertise and time-consuming work. Automating these effortful but effective means to enable bias-sensitive news consumption is the objective of this thesis.

F. Hamborg, *Revealing Media Bias in News Articles*,
https://doi.org/10.1007/978-3-031-17693-7_3

The current chapter proposes *person-oriented framing analysis (PFA)* to address our research question. In the following, we first provide a definition of media bias (Sect. 3.2). Then, we discuss the solution space to tackle media bias (Sect. 3.3). Using the findings of these discussion, we propose the PFA approach to reveal biases in news articles (Sect. 3.4). Lastly, we introduce a side contribution of this thesis, a system for news crawling and extraction (Sect. 3.5). The system collects news articles from online news outlets. The news extractor can be used before PFA to gather articles for analysis and has also demonstrated its usefulness in other use cases throughout the research described in this thesis.

After this chapter, Chaps. 4 and 5 introduce the individual methods part of PFA, and Chap. 6 then introduces our prototype and evaluation to demonstrate the effectiveness of the PFA approach.

3.2   Definition of Media Bias

As our literature shows, many definitions of media bias exist (Sect. 2.2.1). In sum, researchers in the social sciences have proposed various task-specific and in part overlapping or disagreeing definitions of bias. Compared to them, automated approaches tackle bias instead as a single holistic or superficial concept. In the remainder of this thesis, we use a definition of bias reflecting the shared, conceptual understanding established by our literature review.

> **Definition of bias**
> We define *bias* as the effect of framing, i.e., the promotion of "a particular problem definition, causal interpretation, moral evaluation, and/or treatment recommendation" [79], that arises from one or more of the bias forms defined by the news production process.

Bias can exist on various levels due to various means. For example, individual sentences and news articles can be slanted toward a specific attitude due to word choice and labeling, source selection, and other means of bias defined by the news production process (Sect. 2.2.3).

When comparing our definition to the various bias definitions devised in the social sciences, we identify the following commonalities and differences. First, our definition entails both intentional and unintentional bias. Some social science studies distinguish whether bias is intentionally implemented or unintentionally "exists" (cf. [327, 382]). Second, to allow for timely identification of bias, our definition allows bias to emerge from single incidents. In contrast, researchers in the social sciences analyze bias as a systematic tendency, i.e., an effect of multiple observations on extended time frames, since they are typically interested in the effects or implications of (biased) coverage, e.g., on society or policy decisions (cf. [382]).

Third, our definition is task-specific. Identical to social science research, which specific forms of bias are analyzed and how depends on the task at hand and research question. For example, in the social sciences, researchers devise features, such as frames, which they then quantify in the data, e.g., using content analysis. In our definition, the analyzed features are frames due to specific bias forms as described later in this chapter (Sect. 3.3.2). Fourth, identical to social science research, our definition is fundamentally based on the relativity of bias. Bias can only emerge from comparing multiple pieces of information (cf. [78, 288], Sect. 2.3). Newsreaders may have the sensation of bias (whether factually founded or not) when comparing news items with another or with their own attitude. Social science researchers compare news articles, e.g., with another, over a time frame, or with other information sources, such as police reports.

3.3 Discussion of the Solution Space

As our literature review shows, the spectrum of means to tackle media bias is as diverse as the complex issue of media bias itself. This section discusses questions that guide us toward a specific solution to address our research question. We summarize the findings of our literature review and discuss them in the context of our research question. Section 3.3.1 discusses when media bias can be tackled and the broad spectrum of means to tackle bias. Section 3.3.2 discusses approaches to address our research question specifically. Both discussions also strengthen the brief reasoning of our research question from Chap. 1.

3.3.1 Tackling Media Bias

Before discussing *how* media bias can be tackled, we need to discuss *when* it can be tackled.

Tackling Media Bias During News Production or After

Park et al. [276] distinguish two cases to tackle bias: during the production of news and afterward. The various means during the news production aim to prevent media bias in the first place or exposing it explicitly. Such means range from setting the goal to write and publish "objective" news coverage [381] to news formats that contrast media perspectives (so-called press reviews) or that are intended to explicitly convey the journalist's opinion, such as columns, commentaries, or reviews.

According to Park et al. [276], all such means are impractical or inefficient. For example, defining "objective" news coverage is difficult or even impossible in a meaningful way. This can, for example, trivially be seen when looking at bias by event election. Given the myriads of events happening every day, journalists *have* to select a tiny subset to report on. Albeit necessary, how could this event selection be objective? Even when allowing "some degree of tolerable bias" [382], the fundamental issue remains: any approach aiming at objective—or tolerably biased—coverage will fail as long as there is no objectively measurable definition of such coverage.

There are various other reasons why news is rather biased than not. Perhaps most importantly, the news is meant to put events into context and assess the events' meaning for individuals and society (see Chap. 2). Press standards, for example, often do not set objectivity as a higher-level goal but instead fairness or human dignity [360]. News publishers have also a financial incentive to at least slightly slanting their coverage toward the ideology of their target audience (see Sect. 2.2.3).

Lastly, bias-sensitive news formats, such as press reviews or commentaries, are an interesting means to tackle bias but suffer from the following issues. For example, press review can naturally only be created *after* other event coverage was already published. Similarly, a commentary can only serve as a source for an expert's opinions and thoughts regarding an issue. While valuable as such, commentaries are by definition far from factual reporting. Thus, neither of such news formats can be a primary or universal form of news coverage. Instead, these forms can only complement up-to-date news formats, such as event coverage.

We conclude that avoiding slanted coverage or generally tackling media bias during news production seems infeasible or at least impractical. Because of the term's vagueness, defining the term "objective" is problematic in the first place, as is adhering to it if set as a goal for news production. Press standards and research from the social sciences suggest that biases are structurally inherent to news coverage. In principle, a diversity of opinions and slants in the news can even considered to be desirable. Consequently, investigating post-production means to tackle media bias seems more suitable concerning our research question and also in general.

Post-production Means to Tackle Biases After the Production of News

We identify three conceptual categories of means to address media bias: bias analysis, bias correction, and bias communication.[1] The first category, *bias analysis*,

---

[1] Our three categories are in part adapted from the four categories by Park et al. [276] with the following key differences. Our category "bias analysis" matches their categories "bias diagnosis" and "bias measurement." Since both of their categories at their core aim at identifying biases (one category focusing on qualitative analysis, the other on quantitative, rather) [276], we consider them conceptually very similar concerning our research question and summarize them in one category. Further, our "bias communication" is approximately similar to their category "bias mitigation," which in other disciplines, such as psychology [142], medicine [265], and information visualization [372], is partially also called "cognitive bias mitigation." However, these terms differ insofar from another that our term more explicitly highlights the need to communicate bias to mitigate the negative effects of bias, rather than mitigating bias itself. We think this definition better reflects that slanted news coverage is not harmful per se and may even be desirable if readers are aware of the biases. See also Sect. 1.1 and Chap. 2 for a discussion on this matter.

includes both manual and automated approaches to identify and analyze biases. As we discuss in Sect. 2.3, to date, the most effective methods rely on costly manual analysis, e.g., systematic reading and annotation as part of content analysis. Scalable, automated approaches exist but yield less substantial or inconclusive results, especially when comparing their results to manual approaches from the social sciences. For example, they focus on quantitative properties, thereby missing the "meaning between the lines." Or they analyze only vaguely defined instances of media bias, such as "topic diversity" [252] or "differences of [news] coverage" [278] (Sect. 2.2.4). In practical terms, automated approaches find technically significant biases, which, however, are often not meaningful or do not represent all frames of an event's coverage.

Second, approaches for *bias correction* aim to identify biases and then "correct" [276] them, e.g., by removing biases or replacing biased statements with (more) neutral statements conveying the same information. The category represents a relatively young line of computer science research. Bias correction lacks an equivalent in social science research on media bias, possibly because of the previously mentioned characteristics (biases are inherent to the news and may—in principle— even be desirable to facilitate a rich diversity in opinions). The few automated approaches are mostly exploratory and yield mixed results [276]. For example, one recent approach aims to identify and then flip the slant of news headlines, e.g., from having a left stance to become right-slanted [49]. The poor results indicate the complexity of this task. 63% of the generated headlines with flipped slant were not even understandable. In only 42%, the bias could be flipped while still reporting on the initial headline's event. Other approaches rely heavily on user-provided feedback (cf. [276]). Here, the lack of a ground truth comparison can be very problematic since users bring their own biases. The two fundamental issues of current approaches for bias correction are as follows. First, defining unbiased news is practically impossible as stated previously [276]. Second, current natural language generation methods do not suffice to reliably produce "corrected" texts from biased news texts, e.g., due to the lack of training datasets (cf. [35, 214]).

Lastly, approaches for *bias communication* aim to inform news consumers about biases, e.g., by showing different slants present in news coverage on a given event. Previous studies find that bias-sensitive visualizations can effectively communicate biases and help news consumers to become aware of these biases. For example, users of NewsCube's bias-sensitive visualizations read more articles than users of a bias-agnostic baseline. Thereby, the users actively exposed themselves to more diverse perspectives because many articles conveyed perspectives not aligning with the individual users' ideology. In sum, the users of bias-sensitive visualizations developed "more balanced views" [276] on the news events. Similarly, the evaluation of our matrix-based news aggregation finds that users exposed to bias-sensitive visualizations more effectively and more efficiently became aware of the various perspectives present in the news coverage [128]. Besides such academic efforts, other approaches exist to communicate biases during everyday news consumption. For example, AllSides is a bias-aware news aggregator that shows for each topic

one article from a left-wing, center, and right-wing news outlet, respectively [8].
In contrast to popular news aggregators, this approach facilitates showing diverse
perspectives.

In addition to the confirmed effectiveness, studies concerned with bias communication found positive effects on individuals and society. For example, bias
communication supports news consumers in making more informed choices, e.g.,
in elections [22]. However, despite their effectiveness, effectively communicating
biases suffers from the effort of manual techniques for bias identification or the
superficial results yielded by automated approaches.

In sum, we conclude that devising a post-production approach for bias identification (category "bias analysis") and subsequent communication ("bias communication") is the most promising research direction to address the issues caused by
media bias. The previous discussion thus also strengthens the brief reasoning for
our research question described in Chap. 1.

3.3.2   Addressing Our Research Question

One key finding of our literature review is that interdisciplinary research can
improve the effectiveness and efficiency of prior work conducted separately in each
discipline (see Sect. 2.5). The largely manual methods from the social sciences are
effective, e.g., they yield substantial results. However, they are also not as efficient
compared to automated approaches. Simultaneously, while automated approaches
are highly efficient, they are often not as effective as methods employed in social
science research. One reason for the often only superficial or inconclusive results
is the discrepancy in how bias is defined and analyzed in automated approaches
compared to the practice-proven models established in the social sciences.

We aim to combine the relevant methodologies of both disciplines in this thesis.
In particular, we propose an automated approach that roughly resembles the manual
process of frame analysis established in bias research in the social sciences. By
following social science methodology, we can address the previously mentioned
discrepancy that is a fundamental cause for the comparably low performance of
automated approaches. The automated approaches prevalently analyze media bias
as a single holistic or vaguely defined concept (Sect. 1.2). In contrast, the news
production process (Sect. 2.2.3) defines nine strongly different forms of media bias,
each due to strongly different causes and each causing effects on different objects.
Not distinguishing the individual forms and analyzing just one holistic "bias" must
lead to superficial, unmeaningful, or inconclusive results.

So, instead of analyzing vaguely defined biases, such as "subtle differences"
[210], we seek to identify meaningful frames in order to reveal biases. Following
our definition of media bias (Sect. 3.2), we seek to identify substantial frames
by analyzing specific forms defined by the news production process described in
Sect. 2.2.3.

When selecting which forms to identify, we need to balance two goals: representativeness, i.e., covering a broad range of bias forms, and low cost, i.e., covering only a few forms. Our literature review shows that in-depth analysis of individual forms causes exacting effort to achieve substantial and reliable results. Of course, an automated approach would spare much or all of the repetitive effort caused by manual analyses. However, devising a reliable approach for analyzing a particular form still causes high cost initially. For example, because of the forms' differing characteristics, individual methods would need to be devised for each form, each requiring also the creation of a sufficiently large, high-quality dataset for training or at least testing. Thus, on the one hand, focusing on a subset of bias forms seems more feasible than devising analysis methods for all forms individually. On the other hand, focusing on too few forms may cause the approach to miss relevant means of bias in a given news article or coverage. Thus, we aim to cover a sufficiently large set of impactful bias forms while maintaining high specificity and effectiveness through focusing on a set as small as possible. We expect that a well-balanced trade-off between both goals will allow us to identify substantial and meaningful frames.

We propose to identify a fundamental effect resulting from multiple bias forms emerging at the text level rather than analyzing them individually: *effects of person-targeting framing*, i.e., how individual persons are portrayed in the news. Person-targeting framing yields *person-oriented frames*, which roughly resemble the political frames proposed by [79] and used in our definition of bias (Sect. 3.2). However, our person-oriented frames are somewhat exploratory, e.g., implicitly defined and loosely structured. Person-oriented frames emerge, in particular, from the following bias form.[2]

- **Word choice and labeling:** how the word choice affects the perception of individual persons, e.g., due to how a text describes a person, actions performed by the person, or causes of these actions.

More indirectly, person-oriented frames also emerge from the following two forms of media bias.

- **Source selection:** which sources are used when writing a news article and how their content and language affect the perception of individual persons (see Sect. 2.3.2).
- **Commission and omission of information:** which information, such as actions and causes thereof, is included in the article (or left out) from these sources and how this affects the portrayal of individual persons (see Sect. 2.3.3).

---

[2] As stated in Sect. 2.2, bias forms by definition overlap with another. The three forms here can overlap with another as well as with other forms not explicitly mentioned here, such as spin bias (see Sect. 2.3.8).

3.3.3 Research Objective

As a conclusion of the discussion in Sect. 3.3, we define the following research
objective, which we seek to address in this thesis:

> **Devise an approach to reveal substantial biases in English news arti-
> cles reporting on a given political event by automatically identifying
> text-based, person-oriented frames and then communicating them to
> non-expert news consumers. Implement and evaluate the approach and
> its methods.**

Focusing our research objective on person-targeting framing logically misses
biases not related to individual persons. However, persons are especially important
in news articles reporting on policy topics, e.g., because decisions are made
by politicians and affect individuals in society. Further, according to the news
production process (Sect. 2.2.3), the three bias forms jointly represent all means
on the text level to directly affect the perception of persons. Thus, we hypothesize
that focusing our research on the identification of person-targeting framing has
high potential to effectively identify and communicate a significant share of the
biases real-world news coverage consists of. We investigate this hypothesis in our
prototype evaluation (Sect. 6.7).

3.4 Overview of the Approach

We propose *person-oriented framing analysis (PFA)*, an approach to reveal biases
by identifying and communicating how persons are portrayed in individual news
articles. The PFA approach identifies person-targeting forms of bias, most importantly word choice and labeling, source selection, and commission and omission of
information.

This section gives a brief conceptual overview of the analysis workflow and the
individual methods. Chapters 4 and 5 then detail the respective methods and evaluate
them individually. Chapter 6 introduces our prototype system that integrates the
individual methods and subsequently reveals media bias to news consumers.
Chapter 6 also presents our large-scale user study findings, demonstrating the
effectiveness of the PFA approach in increasing bias-awareness in non-expert news
consumers.

Our analysis seeks to find articles that similarly frame the persons involved
in given political event coverage. As shown in Fig. 3.1, the analysis consists of
three components: preprocessing, target concept analysis, and frame analysis. The
analysis takes as an input a set of articles reporting on the same political event
and first performs natural language preprocessing. Secondly, target concept analysis
aims to find which persons occur in the event and identify each person's mentions
across all news articles. Thirdly, automated frame analysis aims to identify how
each article portrays the individual persons, both at the article and sentence levels.

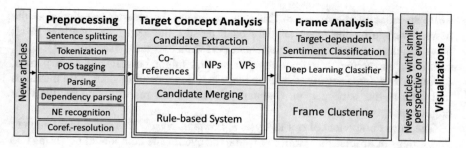

Fig. 3.1 Shown is the three-component analysis workflow as it preprocesses news articles, extracts and resolves phrases referring to the same persons, and groups articles reporting similarly on these persons. Afterward, users can view the analysis results using our visualizations. Adapted from [123]

Afterward, frame analysis clusters those articles that similarly portray the persons involved in the event. The output of our analysis is thus the set of news articles enriched with:

- the set of persons that occur in the news coverage on the event,
- for each such person, all of its mentions resolved across the set of articles,
- for each such mention, weighted framing categories representing how the local context of the mention portrays the person,
- further information derived from the former types of information, including groups of articles with similar perspectives.

In the following, we briefly present each component involved in our analysis workflow. The subsequent chapters of this thesis then provide methodological details.

The input to PFA is a set of news articles written in English reporting on a single event related to politics. We define an *event* as something that happens at a specific and, more importantly, single point in time, typically at a single (geographic) location (cf. [201]). In contrast, we refer to a topic (also called issue) broadly as the "subject of a discourse" [233]. In the context of news coverage, a topic may consist of multiple news events. For example, a news topic might be the 2020 US presidential election. An individual event related to this topic (and of course related to also other topics) is the storming of the US Capitol on January 6, 2021.

The first component of PFA is *preprocessing*. Downstream analysis components, i.e., methods in target concept analysis and frame analysis, use the information extracted during preprocessing. Our preprocessing includes part-of-speech (POS) tagging, dependency parsing, full parsing, named entity recognition (NER), and coreference resolution [56, 57]. We use Stanford CoreNLP with neural models where available, otherwise using the defaults for the English language [224]. Section 4.3.3.1 details our preprocessing.

The second component of PFA is *target concept analysis*. Its objective is to identify which persons are mentioned in the news articles passed to the analysis.

More specifically, the output of this component are the persons mentioned in the news coverage on the event and, for each person, the set of its mentions in all news articles. Therefore, the component performs two tasks (see Fig. 3.1). First, *candidate extraction* to extract any phrase that might be referring to a person. Second, *candidate merging* to resolve these individual mentions, i.e., find mentions that refer to the same person. Chapter 4 describes our research and methods for target concept analysis.

The third component of our analysis is *frame analysis*. This component aims to find groups of articles that similarly frame the event. The component performs two tasks to identify the framing. First, frame analysis determines how each news article portrays the individual persons identified earlier. More specifically, frame analysis determines for each mention how the mention's local context, e.g., the surrounding sentence, portrays the person referred by the mention. Chapter 5 details our research and methods for this part of the frame analysis component. Second, frame analysis uses clustering techniques so that articles similarly portraying the individual persons are part of the same framing group (Sect. 6.3).

Lastly, our prototype system for bias identification and communication reveals the identified framing groups of articles. We devise visualizations intended to aid in the typical news consumption workflow, i.e., first to get an overview of current events and second to get more details on an event, e.g., by reading one or more individual news articles reporting on it. Chapter 6 details our prototype, visualizations, and large-scale user study to demonstrate the effectiveness of the PFA approach.

In addition to these core components of our analysis, we provide an optional component for data gathering, which can be used before the analysis to collect relevant news articles conveniently. Specifically, we present a *web crawler and extractor* for news articles. The system named news-please takes, for example, a set of URLs pointing to article web pages and extracts structured information, such as title, lead paragraph, and main text. Subsequently, this information can be passed to the system. The following section describes the crawler and extractor in more detail.

3.5　Before the Approach: Gathering News Articles

This section details our method and system for crawling and extracting news articles from online news outlets. Besides the need for such a system in this thesis, e.g., to conveniently acquire news coverage on a specific event, there is also a general need in the research community that motivated devising the system. For example, while news datasets such as RCV1 [205] are freely available, researchers often need to compile their own dataset, e.g., to include news published by specific outlets or in a certain time frame. Due to the lack of a publicly available, integrated crawler and extractor for news, researchers often implement such tools redundantly. The process of gathering news data typically consists of two phases: (1) *crawling news websites* and (2) *extracting information from news articles*.

Crawling news websites can be achieved using many web crawling frameworks, such as *scrapy* for Python [188]. Such frameworks traverse the links of websites, hence need to be tailored to the specific use case.

Extracting information from news articles is required to convert the raw data that the crawler retrieves into a format that is suitable for further analysis tasks, such as natural language processing. Information to be extracted typically includes the headline, authors, and main text. *Website-specific extractors*, such as used in [234, 270], must be tailored to the individual websites of interest. These systems typically achieve high precision and recall for their extraction task, but require significant initial setup effort in order to customize the extractors to a set of specific news websites. Such website-specific extractors are most suitable when high data quality is essential, but the number of different websites to process is low.

Generic extractors are intended to obtain information from different websites without the need for adaption. They use heuristics, such as link density and word count, to identify the information to be extracted. Our literature review and experiments show that *Newspaper* [396] is currently one of the most sophisticated and best-performing news extractors. It features robust extraction of all major news article elements and supports more than ten languages. Newspaper includes basic crawling, but lacks full website extraction, auto-extraction of new articles, and news content verification, i.e., determining whether a page contains a news article. The extraction performance of other frameworks, such as *boilerpipe* [185], *Goose* [185], and *readability* [114], is lower than that of the *Newspaper* tool. Furthermore, these latter tools do not offer support for crawling websites.

To our knowledge, no available tool fully covers both the crawling and extraction phase for news data. Web crawler frameworks require use case-specific adaptions. News extractors lack comprehensive crawling functionality. Existing systems lack several key features, particularly the capability (1) to extract information from *all* articles published by a news outlet (*full website extraction*) and (2) to auto-extract newly published articles. With *news-please*, we provide a system that addresses these two weaknesses using a generic crawling and extraction approach. The following section details the processing pipeline of news-please.

3.5.1 Method

news-please is a news crawler and extractor developed to meet five requirements: (1) broad coverage—extract news from any outlet's website; (2) full website extraction; (3) high quality of extracted information; (4) ease of use, simple initial configuration; and (5) maintainability. Where possible, news-please combines prior tools and methods, which we extended with functionality to meet the outlined requirements. This section describes the processing pipeline of news-please as shown in Fig. 3.2.

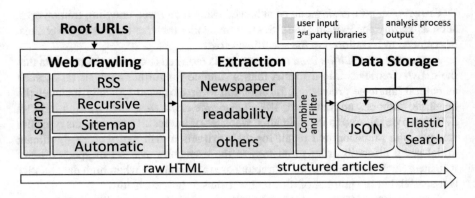

Fig. 3.2 Pipeline for news crawling and extraction. Source [136]

Root URLs Users provide URLs that point to the *root* of news outlets' websites, e.g., https://nytimes.com/. For each root URL, the following tasks are performed.

Web Crawling news-please performs two sub-tasks in this phase. (1) The *crawler* downloads articles' HTML, using the *scrapy* framework. (2) To find all articles published by the news outlet, the system supports four techniques: *RSS* (analyzing RSS feeds for recent articles), *recursive* (following internal links in crawled pages), *sitemap* (analyzing sitemaps for links to all articles), and *automatic* (tries sitemaps and falls back to recursive in the case of an error). The approaches can also be combined, e.g., by starting two news-please instances in parallel, one in automatic mode to get all articles published so far and another instance in RSS mode to retrieve recent articles.

Extraction We use multiple news extractors to obtain the desired information, i.e., title, lead paragraph, main content, author, date, main image, and language. In preliminary tests (see Sect. 3.5.2), we evaluated the performance of four extractors (*boilerpipe*, *Goose*, *Newspaper*, and *readability*). *Newspaper* yielded the highest extraction accuracy for all news elements combined followed by *readability*. Thus, we integrated both extractors into *news-please*. Because both *Newspaper* and *readability* performed poorly for extracting publication dates, we employ a regex-based date extractor [101]. Because none of the extractors is able to determine the language an article is written in, we employ a library for language detection [66]. Our component-based design allows easily adding or removing extractors in the future. Currently, news-please combines the results of the extractors using rule-based heuristics. We discard pages that are likely not articles using a set of heuristics, such as link-to-headline ratio, and metadata filters.

Data Storage news-please currently supports writing the extracted data to JSON files and to an Elasticsearch interface.

Besides the crawling and extraction workflow outlined previously, news-please supports two further use cases. First, users can directly use the extraction functionality (without the crawling part) for single URLs directly pointing to individual online news articles to retrieve the articles' content as structured information, e.g., consisting of title and the other categories outlined previously. Second, news-please allows users to conveniently access the Common Crawl News Archive [256], which consists as of writing of over 400M potential articles gathered from more 50k potential news sources.[3] Especially the extraction functionality for Common Crawl has been used frequently during the individual research parts summarized in this thesis (see Sect. 3.5.3).

3.5.2   Evaluation

We conducted a preliminary, quantitative evaluation where we asked four assessors (students in computer science, aged between 19 and 25, three male, one female) to rate the quality of the information extracted by news-please and the four approaches described in Sect. 3.5, i.e., Newspaper, readability, Goose, and boilerpipe. We selected 20 articles from 20 news websites (the top 15 news outlets by global circulation and 5 major outlets in Germany) and manually assessed the quality of extracted information using a 4-point Likert scale. Our multi-graded relevance assessment includes the four categories: (A) perfect; (B) good, the beginning of an element is extracted correctly, later information is missing, or information from other elements is wrongfully added; (C) poor, in addition to (B), the beginning of an element is not extracted entirely correctly; and (D) unusable, much information is missing or from other elements. After two to three training iterations with the individual assessors, each including a discussion of their previous annotations, the inter-rater reliability measured with mean pairwise agreement was sufficiently high IRR = 0.78 (measured using average pairwise percentage agreement).

Table 3.1 shows the *mean average generalized precision* (MAgP), a score suitable for multi-graded relevance assessments [168]. The MAgP of news-please was 70.6 over all dimensions, assessors, and articles. Moreover, news-please yielded best extraction performances both overall and for each individual extraction dimension, except for main image and author. It performed particularly well for titles (MAgP = 82.0), description (70.0), date (70.0), and main image (76.0). For main content (63.6) and author (30.3), performance was worse but still better than or similar to the other approaches. Overall, news-please performed better than the included extractors individually.

---

[3] Since the archive was created automatically, many of its data items are in fact not news articles and even less report on policy issues (cf. [96]). Yet, to our knowledge, the Common Crawl News Archive represents the largest archive of news articles to date.

Table 3.1 MAgP-performance of news-please and other news extractors. "Desc." refers to the description, i.e., the lead paragraph, "Img." to the article's main image, and "Lang." to the language the article is written in

Approach	Title	Desc.	Text	Author	Date	Img.	Lang.
newspaper	80.3	33.0	57.0	**31.0**	61.0	**89.0**	0.00
readability	42.6	37.0	**64.8**	0.00	22.2	55.6	0.00
goose	66.7	51.9	24.1	0.00	0.00	0.00	0.00
boilerpipe	59.1	44.0	36.8	16.5	55.1	0.00	0.00
news-please	**88.0**	**70.0**	63.6	**30.3**	**70.0**	76.0	**94.0**

3.5.3   Conclusion

We presented *news-please*, the first integrated crawler and information extractor designed explicitly for news articles. The system is designed to be able to crawl *all* articles of a news outlet, including articles published during the crawling process. The system combines the results of three state-of-the-art extractors. For high maintainability and extendibility, news-please allows for inclusion of additional extractors and adaption to use case-specific requirements, e.g., by adding an SQL result writer.

Our quantitative evaluation with four assessors found that news-please overall achieves a higher extraction quality than the individual extractors. By integrating both the crawling and extraction task, researchers can gather news faster and with less initial and long-term effort.

Within the context of this thesis, the system provides a convenient way to collect news articles that can then be analyzed for bias and subsequently visualized. We find that the system effectively helps to reduce the amount of manual work required throughout many use cases of this thesis, i.e., during the creation of our datasets for event detection (Sect. 4.2), coreference resolution and frame properties (Sect. 4.3), target-dependent sentiment classification (Chap. 5), and, finally, the user study (Chap. 6). In all cases, we use news-please to gather news articles and extract structured information from them, which we then manually revise for extraction errors. Due to the on average high extraction performance, we find that the amount of manual work required for revising the data is much lower than for manually extracting the news articles' data from their respective web pages. Other researchers have used the output of news-please without manual verification. For example, Liu et al. [214] used our system to create part of their large-scale dataset to pre-train the widespread deep language model *RoBERTa*.

The system and code are available at https://github.com/fhamborg/news-please.

3.6 Summary of the Chapter

This chapter proposed person-oriented framing analysis (PFA), our approach to reveal biases in news articles. By discussing the findings of our literature review in the context of our research question, we narrowed down our intentionally broadly defined and open research question to a specific research objective. Specifically, we concluded that of the three conceptual means to address media bias, the following two are the most promising to address our research question effectively: *bias analysis*, which aims to identify biases present in news coverage, and *bias communication*, which aims to inform news consumers about such biases.

Methodologically, we narrowed down our research question to identify person-oriented frames as the effects of person-targeting bias forms, especially bias by source selection, commission and omission of information, and word choice and labeling. According to the news production process, the three bias forms jointly represent all means on the text level to affect a person's portrayal directly. Focusing on persons seems promising since news coverage on policy issues is fundamentally about persons, such as individuals in society affected by political decisions or politicians making such decisions. Thus, we hypothesize that our design can cover a wide range of substantial biases while avoiding the issues if we were to analyze all bias forms, e.g., infeasibly high annotation cost or unreliable methods. In Chap. 6, we will investigate the strengths and limitations due to these design decisions.

In sum, the PFA approach takes a set of news articles written in English reporting on the same policy event. The analysis employed by PFA consists of three components. First, we employ *preprocessing*. Second, we perform *target concept analysis* to identify and resolve persons mentioned in the news articles. Third, we perform automated *frame analysis* to identify how each news article portrays the individual persons, also on the sentence level.

A side contribution of this chapter is a system for news crawling and extraction, designed to conveniently gather news articles, such as to be analyzed subsequently using the PFA approach. We will also use this news extractor in all parts of this thesis to create our training and test datasets (see Sect. 3.5.3).

In the following chapters, we will introduce the individual analysis components of PFA. For each analysis component, we will explore different methods to tackle the respective component's goals. Afterward, we will demonstrate the effectiveness of the PFA approach in a large-scale user study.

Chapter 4
Target Concept Analysis

Abstract This chapter details the first component of person-oriented framing analysis: target concept analysis. This component aims to find and resolve mentions of persons, which can be subject to media bias. The chapter introduces and discusses two approaches for this task. First, the chapter introduces an approach for event extraction. The approach extracts answers to the journalistic 5W1H questions, i.e., who did what, when, where, why, and how. The in-text answers to these questions describe a news article's main event. Afterward, the chapter introduces an approach that is the first to resolve highly context-dependent coreferences across news articles as they commonly occur in the presence of sentence-level bias forms. Our approach can resolve mentions that are coreferential also only in coverage on the same event and that otherwise may even be contradictory, such as "attack" or "self-defense" and "riot" or "protest." Lastly, the chapter argues for using the latter approach for the target concept analysis component, in particular because of its high classification performance. Another reason for our decision is that using the event extraction approach in the target concept analysis component would require the development of a subsequent approach, i.e., to compare the events extracted from individual articles and resolve them across all articles.

4.1 Introduction

Target concept analysis finds and resolves mentions of persons that can be subject to the bias forms analyzed by person-oriented framing analysis (PFA), in particular, word choice and labeling, source selection, and commission and omission of information (Sect. 3.3.2). This task is of particular importance in PFA and difficulty in slanted coverage, for example, due to the divergent word choice across differently slanted news articles. While one article might refer to "undocumented immigrants," others may refer to "illegal aliens." Also, within single articles, different terms may be used to refer to the same persons, such as referring to "Kim Jong-un" and later quoting a politician using the term "little rocket man" [74].

In PFA, target concept analysis is the first analysis component (Fig. 3.1). The input to the target concept analysis is a set of news articles reporting on the same

© The Author(s) 2023
F. Hamborg, *Revealing Media Bias in News Articles*,
https://doi.org/10.1007/978-3-031-17693-7_4

event. The output should be the set of all persons mentioned in the event coverage and each person's mentions resolved across all news articles.

We investigate two conceptually different approaches to tackle the task of target concept analysis. Section 4.2 describes our first approach, *main event extraction*. The approach extracts phrases describing a given article's main event, e.g., who is the main actor and what action is performed by the main actor.

Section 4.3 describes our method for *context-driven cross-document coreference resolution*, which is a technique to find mentions of semantic concepts, such as persons, and resolve them across one or more text documents, i.e., news articles in the context of our thesis. Compared to event extraction, coreference resolution allows for directly finding and resolving all mentions of persons (and other concept types) across the given news articles.

Lastly, Sect. 4.4 contrasts both approaches and reasons why coreference resolution is used as the primary approach employed in target concept analysis.

4.2 Event Extraction

Methods in the field of event extraction aim to determine one or more events in a given document and extract specific properties of these events [388]. In the context of target concept analysis, event extraction can be useful as the first step of two. In the first step, we would use event extraction to extract phrases describing important properties of each article's events, such as the actor, which action the actor performed, where, when, and to whom. In the second step, we would then analyze across the articles which events are identical. Using this way matched events, we could deduce that their individual properties refer to the same concepts. At the end of this process, mentions, even those apparently dissimilar, could be resolved across the set of articles. Table 4.1 shows a simple example of the previously outlined idea, where—for the sake of simplicity only two—event properties, i.e., the actor ("who") and action ("what") performed by the actor, are extracted from articles' sentences and headlines (first column). Despite the textual and semantic difference of "illegal aliens" and "undocumented immigrants," our approach could resolve them to the same semantic concept, here a group of persons, due to the action's similarity, which is in both cases "cross [the] border."

Table 4.1 Simplistic example showing two of the six 5W1H properties where the "what" properties are semantically identical

Sentence	Who	What
Illegal aliens cross the border.	illegal aliens	cross the border
Undocumented immigrants cross border	undocumented immigrants	cross border

**Taliban attacks German consulate in northern Afghan city of
Mazar-i-Sharif with truck bomb**

*The death toll from a powerful Taliban truck bombing at the German
consulate in Afghanistan's Mazar-i-Sharif city rose to at least six Friday,
with more than 100 others wounded in a major militant assault.*

The Taliban said the bombing late Thursday, which tore a massive
crater in the road and overturned cars, was a "revenge attack" for US air
strikes this month in the volatile province of Kunduz that left 32 civil-
ians dead. [...] The suicide attacker rammed his explosives-laden car into
the wall [...].

Fig. 4.1 News article consisting of title (bold), lead paragraph (italic), and first of remaining
paragraphs. Highlighted phrases represent the 5W1H event properties (who did what , when ,
where , why , and how). Source [2]

To tackle the first step of the previously outlined idea, we propose *Giveme5W1H*,
a method to extract phrases answering the journalistic 5W1H questions.[1] Figure 4.1
depicts an example of 5W1H phrases, which describe an article's main event, i.e.,
who does *what*, *when*, *where*, *why*, and *how*. We also introduce an annotated dataset
for the evaluation of the approach.

Specifically, our objective is to devise Giveme5W1H as an automated method for
extracting the main event being reported on by a given news article. For this purpose,
we exclude non-event-reporting articles, such as commentaries or press reviews.
First, we define the extracted main event descriptors to be *concise* (requirement
R1). This means they must be as short as possible and contain only the information
describing the event while also being as long as necessary to contain all information
of the event. Second, the descriptors must be of *high accuracy* (*R2*). For this reason,
we give higher priority to extraction accuracy than execution speed.

The remainder of this section is structured as follows. Section 4.2.1 discusses
prior work in event extraction. Section 4.2.2 details our event extractor. Section 4.2.3
presents the results of our evaluation of the system. Section 4.2.4 discusses the
system's performance both concerning general event extraction and concerning
PFA. Section 4.2.5 concludes this line of research and reasons why we focus on
a second line of research for target concept analysis, which we then describe in
Sect. 4.3.

Giveme5W1H and the datasets for training and evaluation are available at
https://github.com/fhamborg/Giveme5W1H.

---

[1] Giveme5W1H represents the recent result of our research on the extraction of main event
descriptors from news articles. An earlier variant, named Giveme5W [133], extracted phrases
answering only the 5W questions and used a simpler methodology, achieving lower extraction
accuracy compared to Giveme5W1H. See also Sect. 4.2.2.

4.2.1   Related Work

This section gives a brief overview of 5W and 5W1H extraction methods in the news domain. Most systems focus only on the extraction of 5W phrases without "how" phrases (cf. [67, 345, 392, 393]). The authors of prior work do not justify this, but we suspect two reasons. First, the "how" question is particularly difficult to extract due to its ambiguity, as we will explain later in this section. Second, "how" (and "why") phrases are considered less important in many use cases when compared to the other phrases, particularly those answering the "who," "what," "when," and "where" (4W) questions (cf. [156, 349, 397]). For the sake of readability in this section, we include approaches to extract both 5W and 5W1H when referring to 5W1H extraction. Aside from the "how" extraction, the analysis of approaches for 5W1H or 5W extraction is conceptually identical.

The task is closely related to closed-domain question answering, which is why some authors call their approaches *5W question answering* (QA) systems.

Systems for 5W QA on news texts typically perform three tasks to determine the article's main event [374, 393]: (1) *preprocessing*; (2) *phrase extraction* [88, 176, 321, 392, 397], where, for instance, linguistic rules are used to extract phrase candidates; and (3) *candidate scoring*, which selects the best answer for each question by employing heuristics, such as the position of a phrase within the document. The input data to QA systems is usually text, such as a full article including headline, lead paragraph, and main text [321], or a single sentence, e.g., in news ticker format [392]. Other systems use automatic speech recognition (ASR) to convert broad casts into text [393]. The outcomes of the process are six textual phrases, one for each of the 5W1H questions, which together describe the main event of a given news text, as highlighted in Fig. 4.1.

The *preprocessing task (1)* performs sentence splitting, tokenizes them, and often applies further NLP methods, including part-of-speech (POS) tagging, coreference resolution [321], NER [88], parsing [225], or semantic role labeling (SRL) [47].

For the *phrase extraction task (2)*, various strategies are available. Most systems use manually created linguistic rules to extract phrase candidates from the preprocessed text [176, 321, 393]. Noun phrases (NP) yield candidates for "who," while sibling verb phrases (VP) are candidates for "what" [321]. Other systems use NER to only retrieve phrases that contain named entities, e.g., a person or an organization [88]. Other approaches use SRL to identify the agent ("who") performing the action ("what") and location and temporal information ("where" and "when") [392]. Determining the reason ("why") can even be difficult for humans because often the reason is only described implicitly, if at all [108]. The applied methods range from simple approaches, e.g., looking for explicit markers of causal relations [176], such as "because," to complex approaches, e.g., training machine learning (ML) methods on annotated corpora [11]. The clear majority of research has focused on explicit causal relations, while only few approaches address implicit causal relations, which also achieve lower precision than methods for explicit causes [27].

The *candidate scoring task (3)* estimates the best answer for each 5W question. The reviewed 5W QA systems provide only few details on their scoring. Typical heuristics include shortness of a candidate, as longer candidates may contain too many irrelevant details [321], "who" candidates that contain an NE, and active speech [393]. More complex methods are discussed in various linguistic publications and involve supervised ML [165, 392]. Yaman, Hakkani-Tur, and Tur [392] use three independent subsystems to extract 5W answers. A trained SVM then decides which subsystem is "correct" using features, such as the agreement among subsystems or the number of non-null answers per subsystem.

Both the extraction of phrases answering the "why" and "how" questions pose a particular challenge in comparison to the other questions. Determining the reason or cause (i.e.. "why") can even be difficult for humans. Often the reason is unknown, or it is only described implicitly, if at all [108]. Extracting the "how" answer is also difficult, because this question can be answered in many ways. To find "how" candidates, the system by Sharma et al. [321] extracts the adverb or adverbial phrase within the "what" phrase. The tokens extracted with this simplistic approach *detail the verb*, e.g., "He drove quickly," but do not answer the *method* how the action was performed, e.g., by ramming an explosive-laden car into the consulate (in the example in Fig. 4.1), which is a prepositional phrase. Other approaches employ ML [175], but have not been devised for the English language. In summary, few approaches exist that extract "how" phrases. The reviewed approaches provide no details on their extraction method and achieve poor results, e.g., they extract adverbs rather than the tool or the method by which an action was performed (cf. [157, 175, 321]).

While the evaluations of the reviewed papers generally indicate sufficient quality to be usable for news event extraction, e.g., the system by Yaman, Hakkani-Tur, and Tur [392] achieved macro $F1 = 0.85$ on the Darpa corpus from 2009, they lack comparability for two reasons: (1) There is no gold standard for journalistic 5W1H question answering on news articles. A few datasets exist for automated question answering, specifically for the purpose of disaster tracking [202, 350]. However, these datasets are so specialized to their own use cases that they cannot be applied to the use case of automated journalistic question answering. Another challenge to the evaluation of news event extraction is that the evaluation datasets of previous papers are no longer publicly available [279, 392, 393]. (2) Previous papers use different quality measures, such as precision and recall [67] or error rates [393].

Another weakness of the reviewed prior work is that none of them yield *canonical* or *normalized data*. Canonical output is more concise and also less ambiguous than its original textual form (cf. [378]), e.g., polysemes, such as crane (animal or machine), have multiple meanings. Hence, canonical data is often more useful in downstream analyses (see Sect. 4.2). Phrases containing temporal information or location information may be canonicalized, e.g., by converting the phrases to dates or timespans [48, 343] or to precise geographic positions [207]. Phrases answering the other questions could be canonicalized by employing NERD on the contained NEs and then linking the NEs to concepts defined in a knowledge graph, such as YAGO [150] or WordNet [239].

In sum, methods for extracting events from articles suffer from three main short-comings. First, most approaches only detect events implicitly, e.g., by employing topic modeling [90, 355]. Second, they are specialized for the extraction of task-specific properties, e.g., extracting only the number of injured people in an attack [267, 355]. Lastly, some methods extract explicit descriptors, but are not publicly available, or are described in insufficient detail to allow researchers to reimplement the approaches [279, 374, 392, 393].

4.2.2 Method

Giveme5W1H is a method for main event retrieval from news articles that addresses the objectives we defined in Sect. 4.2. The system extracts 5W1H phrases that describe the most defining characteristics of a news event, i.e., *who* did *what*, *when*, *where*, *why*, and *how*. This section describes the analysis workflow of Giveme5W1H, as shown in Fig. 4.2. Due to the lack of a large-scale dataset for 5W1H extraction (see Sect. 4.2.1), we devise the system using traditional machine learning techniques and domain knowledge.

Besides the intended use in PFA, Giveme5W1H can be accessed by other software as a Python library and via a RESTful API. Due to its modularity, researchers can efficiently adapt or replace components. For example, researchers can integrate a custom parser or adapt the scoring functions tailored to the characteristics of their data. The system builds on our earlier system, Giveme5W [133], but improves the extraction performance by addressing the planned future work directions: Giveme5W1H uses coreference resolution, question-specific semantic distance measures, combined scoring of candidates, and extracts phrases for the "how" question. The values of the parameters introduced in this section result from a semi-automated search for the optimal configuration of Giveme5W1H using an annotated learning dataset including a manual, qualitative revision (see Sect. 4.2.2.5).

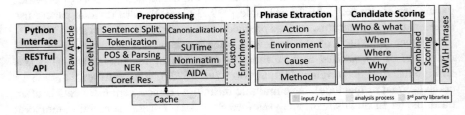

Fig. 4.2 The three-phase analysis pipeline preprocesses a news text, finds candidate phrases for each of the 5W1H questions, and scores these. Giveme5W1H can easily be accessed via Python and via a RESTful API

4.2.2.1   Preprocessing

Giveme5W1H accepts as input the full text of a news article, including headline, lead paragraph, and body text. The user can specify these three components as one or separately. Optionally, the article's publishing date can be provided, which helps Giveme5W1H parse relative dates, such as "yesterday at 1 pm."

During *preprocessing*, we use Stanford CoreNLP for sentence splitting, tokenization, lemmatization, POS tagging, full parsing, NER (with Stanford NER's seven-class model), and pronominal and nominal coreference resolution. Since our main goal is high 5W1H extraction accuracy (rather than fast execution speed), we use the best-performing model for each of the CoreNLP annotators, i.e., the "neural" model if available. We use the default settings for English in all libraries.

After the initial preprocessing, we bring all NEs in the text into their *canonical* form. Following from requirement R1, canonical information is the preferred output of Giveme5W1H, since it is the most concise form. Because Giveme5W1H uses the canonical information to extract and score "when" and "where" candidates, we implement the canonicalization task during preprocessing.

We parse dates written in natural language into canonical dates using SUTime [362]. SUTime looks for NEs of the type date or time and merges adjacent tokens to phrases. SUTime also handles heterogeneous phrases, such as "yesterday at 1 pm," which consist not only of temporal NEs but also other tokens, such as function words. Subsequently, SUTime converts each temporal phrase into a standardized TIMEX3 instance [292]. TIMEX3 defines various types, also including repetitive periods. Since events according to our definition occur at a single point in time, we only retrieve datetimes indicating an exact time, e.g., "yesterday at 6pm," or a duration, e.g., "yesterday," which spans the whole day.

Geocoding is the process of parsing places and addresses written in natural language into canonical *geocodes*, i.e., one or more coordinates referring to a point or area on earth. We look for tokens classified as NEs of the type location (cf. [392]). We merge adjacent tokens of the same NE type within the same sentence constituent, e.g., within the same NP or VP. Similar to temporal phrases, locality phrases are often heterogeneous, i.e., they do not only contain temporal NEs but also function words. Hence, we introduce a locality phrase merge range $r_{where} = 1$, to merge phrases where up to r_{where} arbitrary NE tokens are allowed between two location NEs. Lastly, we geocode the merged phrases with Nominatim,[2] which uses free data from OpenStreetMap.

We canonicalize NEs of the remaining types, e.g., persons and organizations, by linking NEs to concepts in the YAGO graph [221] using AIDA [150]. The YAGO graph is a knowledge base, where nodes in the graph represent semantic concepts that are connected to other nodes through attributes and relations. The data is derived from other well-established knowledge bases, such as Wikipedia, WordNet, Wikidata, and GeoNames [345].

---

[2] https://github.com/openstreetmap/Nominatim, v3.0.0

4.2.2.2   Phrase Extraction

Giveme5W1H performs four independent extraction chains to retrieve the article's main event: (1) the action chain extracts phrases for the "who" and "what" questions, (2) environment for "when" and "where," (3) cause for "why," and (4) method for "how."

The *action extractor* identifies who did what in the article's main event. The main idea for retrieving "who" candidates is to collect the subject of each sentence in the news article. Therefore, we extract the first NP that is a direct child to the sentence in the parse tree and that has a VP as its next right sibling. We discard all NPs that contain a child VP, since such NPs yield lengthy "who" phrases. Take, for instance, this sentence: "*((NP)* Mr. Trump, *((VP)* who stormed to a shock election victory on Wednesday)), *((VP)* said it was [...])*," where "who stormed [...]" is the child VP of the NP. We then put the NPs into the list of "who" candidates. For each "who" candidate, we take the VP that is the next right sibling as the corresponding "what" candidate. To avoid long "what" phrases, we cut VPs after their first child NP, which long VPs usually contain. However, we do not cut the "what" candidate if the VP contains at most $l_{what,min} = 3$ tokens, and the right sibling to the VP's child NP is a prepositional phrase (PP). This way, we avoid short, undescriptive "what" phrases. For instance, in the simplified example, "*((NP)* The microchip) *((VP)* is *((NP)* part) *((PP)* of a wider range of the company's products))*," the truncated VP "is part" contains no descriptive information; hence, our presented rules prevent this truncation.

The *environment extractor* retrieves phrases describing the temporal and locality context of the event. To determine "when" candidates, we take TIMEX3 instances from preprocessing. Similarly, we take the geocodes as "where" candidates.

The *cause extractor* looks for linguistic features indicating a causal relation within a sentence's constituents. We look for three types of cause-effect indicators (cf. [176, 177]): *causal conjunctions*, *causative adverbs*, and *causative verbs*. Causal conjunctions, e.g., "due to," "result of," and "effect of," connect two clauses, whereas the second clause yields the "why" candidate. For causative adverbs, e.g., "therefore," "hence," and "thus," the first clause yields the "why" candidate. If we find that one or more subsequent tokens of a sentence match with one of the tokens adapted from Khoo et al. [176], we take all tokens on the right (causal conjunction) or left side (causative adverb) as the "why" candidate.

Causative verbs, e.g., "activate" and "implicate," are contained in the middle VP of the causative NP-VP-NP pattern, whereas the last NP yields the "why" candidate [108, 177]. For each NP-VP-NP pattern we find in the parse tree, we determine whether the VP is causative. To do this, we extract the VP's verb, retrieve the verb's synonyms from WordNet [239], and compare the verb and its synonyms with the list of causative verbs from Girju [108], which we also extended by their synonyms (cf. [108]). If there is at least one match, we take the last NP of the causative pattern as the "why" candidate. To reduce false positives, we check the NP and VP for the causal constraints for verbs proposed by Girju [108].

The *method extractor* retrieves "how" phrases, i.e., the method by which an action was performed. The combined method consists of two sub-tasks, one analyzing *copulative conjunctions* and the other looking for *adjectives* and *adverbs*. Often, sentences with a copulative conjunction contain a method phrase in the clause that follows the copulative conjunction, e.g., "after [the train came off the tracks]." Therefore, we look for copulative conjunctions compiled from the Oxford English Dictionary [268]. If a token matches, we take the right clause as the "how" candidate. To avoid long phrases, we cut off phrases longer than $l_{\text{how,max}} = 10$ tokens. The second sub-task extracts phrases that consist purely of adjectives or adverbs (cf. [321]), since these often represent how an action was performed. We use this extraction method as a fallback, since we found the copulative conjunction-based extraction too restrictive in many cases.

4.2.2.3 Candidate Scoring

The last task is to determine the best candidate for each 5W1H question. The scoring consists of two sub-tasks. First, we score candidates independently for each of the 5W1H questions. Second, we perform a combined scoring where we adjust scores of candidates of one question dependent on properties, e.g., position, of candidates of other questions. For each question q, we use a scoring function that is composed as a weighted sum of n scoring factors:

$$s_q = \sum_{i=0}^{n-1} w_{\text{q},i} s_{\text{q},i},$$
(4.1)

where $w_{\text{q},i}$ is the weight of the scoring factor $s_{\text{q},i}$.

To score "who" candidates, we define three scoring factors: the candidate shall occur in the article (1) *early* and (2) *often* and (3) contain a *named entity*. The first scoring factor targets the concept of the *inverse pyramid* [52]: news mention the most important information, i.e., the main event, early in the article, e.g., in the headline and lead paragraph, while later paragraphs contain details. However, journalists often use so-called hooks to get the reader's attention without revealing all content of the article [283]. Hence, for each candidate, we also consider the frequency of similar phrases in the article, since the primary actor involved in the main event is likely to be mentioned frequently in the article. Furthermore, if a candidate contains a NE, we will score it higher, since in news, the actors involved in events are often NEs, e.g., politicians. Table 4.2 shows the weights and scoring factors.

To calculate these factors, we define $\text{pos}(c) = 1 - \frac{n_{\text{pos}}(c)}{d_{\text{len}}}$, $f(c) = \frac{n_f(c)}{\max_{c' \in C}(n_f(c'))}$, where n_{pos} is the candidate c's position measured in sentences within the document, $n_f(c)$ the frequency of phrases similar to c in the document, and $\text{NE}(c) = 1$ if c contains an NE, else 0 (cf. [88]). To measure $n_f(c)$ of the actor in candidate c, we use the number of the actor's coreferences, which we extracted during coreference

Table 4.2 Weights and scoring factors for "who" phrases

i Name		$w_{who,i}$	$s_{who,i}$
0	position	0.900	$pos(c)$
1	frequency	0.095	$f(c)$
2	type	0.005	$NE(c)$

Table 4.3 Weights and scoring factors for "when" phrases

i Name		$w_{when,i}$	$s_{when,i}$
0	position	0.24	$pos(c)$
1	frequency	0.16	$f(c)$
2	containment	0.40	$1 - \min(1, \frac{\Delta_s(c, d_{pub})}{e_{max}})$
3	specificity	0.20	$1 - \min(1, \frac{\log s(c) - \log s_{min}}{\log s_{max} - \log s_{min}})$

resolution (see Sect. 4.2.2.1). This allows Giveme5W1H to recognize and count name variations, as well as pronouns. Due to the strong relation between agent and action, we rank VPs according to their NPs' scores. Hence, the most likely VP is the sibling in the parse tree of the most likely NP: $s_{what} = s_{who}$.

We score temporal candidates according to four scoring factors: the candidate shall occur in the article (1) *early* and (2) *often*. It should also be (3) *close to the publishing date* of the article and (4) of a relatively *short duration*. The first two scoring factors have the same motivation as in the scoring of "who" candidates. The idea for the third scoring factor, close to the publishing date, is that events reported on by news articles often occurred on the same day or on the day before the article was published. For example, if a candidate represents a date one or more years in the past before the publishing date of the article, the candidate will achieve the lowest possible score in the third scoring factor. The fourth scoring factor prefers temporal candidates that have a short duration, since events according to our definition happen during a specific point in time with a short duration. We logarithmically normalize the duration factor between 1 minute and 1 month (cf. [397]). The resulting scoring formula for a temporal candidate c is the sum of the weighted scoring factors shown in Table 4.3.

To count $n_f(c)$, we determine two TIMEX3 instances as similar if their start and end dates are at most 24h apart. $\Delta_s(c, d_{pub})$ is the difference in seconds of candidate c and the publication date of the news article d_{pub}, $s(c)$ the duration in seconds of c, and the normalization constants $e_{max} \approx 2.5Ms$ (1 month in seconds), $s_{min} = 60s$, and $s_{max} \approx 31Ms$ (1 year).

The scoring of *location* candidates follows four scoring factors: the candidate shall occur (1) *early* and (2) *often* in the article. It should also be (3) *often geographically contained* in other location candidates and be (4) *specific*. The first two scoring factors have the same motivation as in the scoring of "who" and "when" candidates. The second and third scoring factors aim to (1) find locations that occur often, either by being similar to others or (2) by being contained in other location candidates. The fourth scoring factor favors specific locations, e.g., Berlin, over broader mentions of location, e.g., Germany or Europe. We logarithmically normalize the location specificity between $a_{min} = 225\,m^2$ (a small property's size) and $a_{max} = 530{,}000\,km^2$ (approx. the mean area of all countries [43]). We discuss other scoring options in Sect. 4.2.4. The used weights and scoring factors are shown in Table 4.4. We measure $n_f(c)$, the number of similar mentions of candidate c, by counting how many other candidates have the same Nominatim place ID. We measure $n_e(c)$ by counting how many other candidates are geographically contained within the bounding box of c, where $a(c)$ is the area of the bounding box of c in square meters.

Scoring *causal* candidates was challenging, since it often requires semantic interpretation of the text and simple heuristics may fail [108]. We define two objectives: candidates shall (1) occur *early* in the document, and (2) their *causal type* shall be reliable [177]. The second scoring factor rewards causal types with low ambiguity (cf. [11, 108]), e.g., "because" has a very high likelihood that the subsequent phrase contains a cause [108]. The weighted scoring factors are shown in Table 4.5. The causal type $TC(c) = 1$ if c is extracted due to a causal conjunction, 0.62 if it starts with a causative RB and 0.06 if it contains a causative VB (cf. [176, 177]).

The scoring of *method* candidates uses three simple scoring factors: the candidate shall occur (1) *early* and (2) *often* in the news article, and (3) their *method type* shall be reliable. The weighted scoring factors for method candidates are shown in Table 4.6.

The method type $TM(c) = 1$ if c is extracted because of a copulative conjunction, else 0.41. We determine the number of mentions of a method phrase $n_f(c)$ by the term frequency (including inflected forms) of its most frequent token (cf. [374]).

Table 4.4 Weights and scoring factors for "where" phrases

i Name	$w_{where,i}$	$s_{where,i}$
0 position	0.37	$pos(c)$
1 frequency	0.30	$f(c)$
2 containment	0.30	$\frac{n_e(c)}{\max_{c' \in C}(n_e(c'))}$
3 specificity	0.03	$1 - \min(1, \frac{\log a(c) - \log a_{min}}{\log a_{max} - \log a_{min}})$

Table 4.5 Weights and scoring factors for "why" phrases

i Name	$w_{why,i}$	$s_{why,i}$
0 position	0.56	$pos(c)$
1 type	0.44	$CT(c)$

Table 4.6 Weights and scoring factors for "how" phrases

i Name	$w_{how,i}$	$s_{how,i}$
0 position	0.23	$pos(c)$
1 frequency	0.14	$f(c)$
2 type	0.63	$TM(c)$

The final sub-task in candidate scoring is *combined scoring*, which adjusts scores of candidates of a single 5W1H question depending on the candidates of other questions. To improve the scoring of method candidates, we devise a combined sentence-distance scorer. The assumption is that the method of performing an action should be close to the mention of the action. The resulting equation for a method candidate c given an action candidate a is:

$$s_{\text{who,new}}(c, a) = s_{\text{how}}(c) - w_0 \frac{|n_{\text{pos}}(c) - n_{\text{pos}}(a)|}{d_{\text{len}}}, \tag{4.2}$$

where $w_0 = 1$. Section 4.2.4 describes additional scoring approaches.

4.2.2.4   Output

The highlighted phrases in Fig. 4.1 are candidates extracted by Giveme5W1H for each of the 5W1H event properties of the shown article. Giveme5W1H enriches the returned phrases with additional information that the system extracted for its own analysis or during *custom enrichment*, with which users can integrate their own preprocessing. The additional information for each token is its POS tag, parse tree context, and NE type if applicable. Enriching the tokens with this information increases the efficiency of the overall analysis workflow in which Giveme5W1H may be embedded, since later analysis tasks can reuse the information.

For the temporal phrases and locality phrases, Giveme5W1H also provides their canonical forms, i.e., TIMEX3 instances and geocodes. For the news article shown in Fig. 4.1, the canonical form of the "when" phrase represents the entire day of November 10, 2016. The canonical geocode for the "where" phrase represents the coordinates of the center of the city Mazar-i-Sharif (36°42'30.8"N 67°07'09.7"E), where the bounding box represents the area of the city and further information from OSM, such as a canonical name and place ID, which uniquely identifies the place. Lastly, Giveme5W1H provides linked YAGO concepts [221] for other NEs.

4.2.2.5   Parameter Learning

Determining the best values for the parameters introduced in Sect. 4.2.2, e.g., weights of scoring factors, is a supervised ML problem [162]. Since there is no gold standard for journalistic 5W1H extraction on news (see Sect. 4.2.1), we created an annotated dataset.

The dataset is available in the open-source repository (see Sect. 4.2.5). To facilitate diversity in both content and writing style, we selected 13 major news outlets from the USA and the UK. We sampled 100 articles from the news categories politics, disaster, entertainment, business, and sports for November 6–14, 2016. We crawled the articles (see Sect. 3.5) and manually revised the extracted information to ensure that it was free of extraction errors.

We asked 3 assessors (graduate IT students, aged between 22 and 26, all male) to read each of the 100 news articles and to annotate the single most suitable phrase for each 5W1H question. Finally, for each article and question, we combined the annotations using a set of combination rules, e.g., if all phrases were semantically equal, we selected the most concise phrase, or if there was no agreement between the annotators, we selected each annotator's first phrase, resulting in three semantically diverging but valid phrases. We also manually added a TIMEX3 instance to each "when" annotation, which was used by the error function for "when." The inter-rater reliability was $IRR_{ann} = 81.0$, measured using average pairwise percentage agreement.

We divided the dataset into two subsets for training (80% randomly sampled articles) and testing (20%). To find the optimal parameter values for our extraction method, we used an exhaustive grid search over all possible parameter configurations.[3] For each parameter configuration, we then calculated the mean error (ME) on the training set. To measure the ME of a configuration, we devised three error functions measuring the semantic distance between candidate phrases and annotated phrases. For the textual candidates, i.e., who, what, why, and how, we used the Word Mover's Distance (WMD) [192]. WMD is a generic measure for semantic similarity of two phrases. For "when" candidates, we computed the difference in seconds between candidate and annotation. For "where" candidates, we computed the distance in meters between both coordinates. We linearly normalized all measures.

We then validated the 5% best-performing configurations on the test set and discarded all configurations that yielded a significantly different ME. Finally, we selected the best-performing parameter configuration for each question.

4.2.3   Evaluation

We conducted a survey with 3 assessors (3 graduate IT students, aged between 22 and 26, all male) and a dataset of 120 news articles, which we sampled from the BBC dataset [115]. The dataset contains 24 news articles in each of the following categories: business ("Bus"), entertainment ("Ent"), politics ("Pol"), sport ("Spo"), and tech ("Tec"). We asked the assessors to read one article at a time. After reading each article, we showed the assessors the 5W1H phrases that had been extracted by the system and asked them to judge the relevance of each answer on a 3-point scale: *non-relevant* (if an answer contained no relevant information, score $s = 0$), *partially relevant* (if only part of the answer was relevant or if information was missing, $s = 0.5$), and *relevant* (if the answer was completely relevant without missing information, $s = 1$).

Table 4.7 shows the *mean average generalized precision* (MAgP), a score suitable for multi-graded relevance assessments [168]. MAgP was 73.0 over all

---

[3] The tested parameter values can be found in the open-source repository.

Table 4.7 IRR and
MAgP-performance of
Giveme5W1H. The last row
displays the mean when
evaluated on only the first
four W questions

Question	IRR	Bus	Ent	Pol	Spo	Tec	Avg.
Who	93.0	98.0	88.0	89.0	97.0	90.0	**92.0**
What	88.0	85.0	69.0	89.0	84.0	66.0	79.0
When	89.0	55.0	91.0	79.0	81.0	82.0	78.0
Where	94.0	82.0	63.0	85.0	79.0	80.0	78.0
Why	95.0	48.0	62.0	42.0	45.0	42.0	48.0
How	87.0	63.0	58.0	68.0	51.0	65.0	61.0
Avg. all	91.0	72.0	72.0	75.0	73.0	71.0	**73.0**
Avg. 4W	91.0	80.0	78.0	86.0	85.0	80.0	**82.0**

categories and questions. If only considering the first 4Ws, which the literature considers as sufficient to represent an event (cf. [156, 349, 397]), overall MAgP was 82.0.

Of the few existing approaches capable of extracting phrases that answer all six 5W1H questions (see Sect. 4.2.1), only one publication reported the results of an evaluation: the approach developed by Khodra achieved a precision of 74.0 on Indonesian articles [175]. Others did not conduct any evaluation [321] or only evaluated the extracted "who" and "what" phrases of Japanese news articles [157].

We also investigated the performance of systems that are only capable of extracting 5W phrases. Our system achieves $MAgP_{5W} = 75.0$, which is 5pp. higher than the MAgP of our earlier system Giveme5W [133]. Directly comparing our system to other systems was not possible (cf. [133]): other systems were tested on non-disclosed datasets [279, 392, 393], they were translated from other languages [279], they were devised for different languages [157, 175, 374], or they used different evaluation measures, such as error rates [393] or binary relevance assessments [392], which are both not optimal because of the non-binary relevance of 5W1H answers (cf. [168]). Finally, none of the related systems have been made publicly available or have been described in sufficient detail to enable a reimplementation.

Therefore, a direct comparison of the results and related work was not possible, but we compared the reported evaluation metrics. Compared to the fraction of correct 5W answers by the best system by Parton et al. [279], Giveme5W1H achieves a 12pp. higher $MAgP_{5W}$. The best system by Yaman, Hakkani-Tur, and Tur [392] achieved a precision $P_{5W} = 89.0$, which is 14pp. higher than our $MAgP_{5W}$ and—as a rough approximation of the best achievable precision [152]—surprisingly almost identical to the inter-rater reliability (IRR) of our assessors.

We found that different forms of journalistic presentation in the five news categories of the dataset led to different extraction performance. Politics articles, which yielded the best performance, mostly reported on single events. The performance on sports articles was unexpectedly high, even though they not only report on single events but also are background reports or announcements, for which event detection is more difficult. Determining the "how" in sports articles was difficult ($MAgP_{how} = 51.0$), since often articles implicitly described the method of an event, e.g., how one team won a match, by reporting on multiple key events during the match. Some categories, such as entertainment and tech, achieved lower extraction performances, mainly because they often contained much background information on earlier events and the actors involved.

4.2.4 Future Work

We plan to improve the extraction quality of the "what" question, being one of the important 4W questions. We aim to achieve an extraction performance similar to the performance of the "who" extraction ($MaGP_{who} = 91.0$), since both are very important in event description. In our evaluation, we identified two main issues: (1) joint extraction of optimal "who" candidates with non-optimal "what" candidates and (2) cut-off "what" candidates. In some cases (1), the headline contained a concise "who" phrase, but the "what" phrase did not contain all information, e.g., because it only aimed to catch the reader's interest, a journalistic hook (Sect. 4.2.1). We plan to devise separate extraction methods for both questions. Thereby, we need to ensure that the top candidates of both questions fit to each other, e.g., by verifying that the semantic concept of the answer of each question, e.g., represented by the nouns in the "who" phrase or verbs in the "what" phrase, co-occurs in at least one sentence of the article. In other cases (2), our strategy to avoid too detailed "what" candidates (Sect. 4.2.2.2) cut off the relevant information, e.g., "widespread corruption in the finance ministry has cost it $2m," in which the underlined text was cut off. We will investigate dependency parsing and further syntax rules, e.g., to always include the direct object of a transitive verb.

For "when" and "where" questions, we found that in some cases, an article does not explicitly mention the main event's date or location. The date of an event may be implicitly defined by the reported event, e.g., "in the final of the Canberra Classic." The location may be implicitly defined by the main actor, e.g., "Apple Postpones Release of [...]," which likely happened at the Apple headquarters in Cupertino. Similarly, the proper noun "Stanford University" also defines a location. We plan to investigate how we can use the YAGO concepts, which are linked to NEs, to gather further information regarding the date and location of the main event. If no date can be identified, the publishing date of the article or the day before it might sometimes be a suitable fallback date.

Using the TIMEX3 instances from SUTime is an improvement ($MAgP_{when} = 78.0$) over a first version, where we used dates without a duration ($MAgP_{when} = 72.0$).

The extraction of "why" and "how" phrases was most challenging, which manifests in lower extraction performances compared to the other questions. One reason is that articles often do not explicitly state a single cause or method of an event, but implicitly describe this throughout the article, particularly in sports articles (see Sect. 4.2.3). In such cases, NLP methods are currently not advanced enough to find and abstract or summarize the cause or method (see Sect. 4.2.2.3). However, we plan to improve the extraction accuracy by preventing the system from returning false positives. For instance, in cases where no cause or method could be determined, we plan to introduce a score threshold to prevent the system from outputting candidates with a low score, which are presumably wrong. Currently, the system always outputs a candidate if at least one cause or method was found.

To improve the performance of all textual questions, i.e., who, what, why, and how, we will investigate two approaches. First, we want to improve measuring a candidate's frequency, an important scoring factor in multiple questions (see Sect. 4.2.2.3). We currently use the number of coreferences, which does not include synonymous mentions. We plan to count the number of YAGO concepts that are semantically related to the current candidate. Second, we found that a few top candidates of the four textual questions were semantically correct but only contained a pronoun referring to the more meaningful noun. We plan to add the coreference's original mention to extracted answers.

Section 4.2 outlined a two-task approach within which Giveme5W1H could be used to tackle the goal of target concept analysis, i.e., identifying and resolving mentions of persons. In the first step, Giveme5W1H would extract the 5W1H event properties. In the second step, these could be resolved across all articles, e.g., by deducing that the actors ("who") of two events refer to the same person if the events' actions are identical. In a simplistic example of two sentences "illegal aliens cross the border" and "undocumented immigrants cross border," this two-task approach could resolve both actors, i.e., "illegal aliens" and "undocumented immigrants," to the same semantic concept, here group of persons, since their action is identical (see Table 4.1).

However, the difficulty of this task increases strongly when not only one but multiple event properties are dissimilar. For example, Table 4.8 shows an additional, third sentence, "The migrant caravan invades the country," which has a different actor ("migrant caravan") and a different action ("invades the country"). In a qualitative investigation of the extracted 5W1H phrases, we find that real-world news coverage often has divergent 5W1H phrases, especially in the presence of bias, making the previously mentioned idea to resolve the mentions infeasible. Moreover, since we want to find and resolve not only a single main actor for each of the event's news articles, we would additionally need to extract fine-grained side events at the sentence level. Lastly, PFA focuses on individual persons, whereas the actors of main events extracted by Giveme5W1H can also be groups of persons, countries, and other concept types.[4] Given this issue of using event extraction, i.e., strongly increased complexity in real-world news articles, and the additionally required work to devise methods for extracting fine-grained side events as well as resolving the event descriptors afterward, we choose to focus our research on a different line of research, which we describe in Sect. 4.3.

---

[4] When beginning with the research on event extraction, we initially planned to include in our overall framing analysis also other concept types than only individual persons.

Table 4.8 Simplistic example showcasing the difficulty of resolving phrases when an event property is only ambiguously similar to others ("what" in the third row)

Sentence	Who	What
Illegal aliens cross the border.	illegal aliens	cross the border
Undocumented immigrants cross border	undocumented immigrants	cross border
The migrant caravan invades the country.	migrant caravan	invades the country

4.2.5 Conclusion

In this section, we proposed *Giveme5W1H*, the first open-source system that extracts answers to the journalistic 5W1H questions, i.e., *who* did *what, when, where, why,* and *how,* to describe a news article's main event. The system canonicalizes temporal mentions in the text to standardized TIMEX3 instances, locations to geocoordinates, and other NEs, e.g., persons and organizations, to unique concepts in a knowledge graph. The system uses syntactic and domain-specific rules to extract and score phrases for each 5W1H question. Giveme5W1H achieved a mean average generalized precision (MAgP) of 73.0 on all questions and an MAgP of 82.0 on the first four W questions (who, what, when, and where), which alone can represent an event. Extracting the answers to "why" and "how" performed more poorly, since articles often only imply causes and methods. Answering the 5W1H questions is at the core of understanding any article and thus an essential task in many research efforts that analyze articles. We hope that redundant implementations and non-reproducible evaluations can be avoided with Giveme5W1H as the first universally applicable, modular, and open-source 5W1H extraction system. In addition to benefiting developers and computer scientists, our system especially benefits researchers from the social sciences, for whom automated 5W1H extraction was previously not made accessible.

In the context of this thesis, the event extraction achieved by Giveme5W1H represents the first step of a two-step approach that could be used to tackle target concept analysis. This approach, outlined in more detail in Sect. 4.2, relies on the idea of first extracting events and matching these across articles in order to determine which event properties refer to the same semantic concepts. Due to conceptual issues as described in Sect. 4.2.4, such as high ambiguity when matching events, we focus our research for target concept analysis on cross-document coreference resolution. Taking this approach also has the advantage of directly extracting and resolving mentions in one method and thus decreases the conceptual complexity of target concept analysis.

Giveme5W1H and the datasets for training and evaluation are available at https://github.com/fhamborg/Giveme5W1H.

4.3 Context-Driven Cross-Document Coreference Resolution

Methods in the field of coreference resolution aim to resolve mentions of semantic concepts, such as persons, in a given text document [298]. Context-driven cross-document coreference resolution (CDCDCR) is a special form of coreference resolution with two differences. First, mentions are identified and resolved across multiple documents. Second, mentions can be less strictly related to another but can still be considered coreferential. Such mentions include also those that are typically non-coreferential, such as "White House" and "US President," or are even contradictory in other contexts, such as "activist" and "extremist." This is an extension to regular (cross-document) coreference resolution, which resolves only mentions that have an identity relation, i.e., that are strictly identical, such as "US President" and "Biden" [298].

In the context of our analysis workflow and in particular the target concept analysis, we use context-driven cross-document coreference resolution to find and identify mentions of persons across the set of news articles reporting on the given event. While our overall bias analysis focuses on person-targeting biases only, we devise our method for context-driven cross-document coreference resolution in this section to resolve also other types of semantic concepts, such as countries, so that the method can be used outside the scope of our analysis.

The remainder of this section is structured as follows.[5] Section 4.3.1 describes related work, highlighting that most research focuses on coreference resolution but only few works exists for cross-document coreference resolution or that aim to resolve mentions with less strictly identical relations. We then describe in Sect. 4.3.2 how we create and annotate our test dataset named *NewsWCL50*. We annotate not only coreferential mentions but also so-called frame properties, e.g., how the persons we annotate are portrayed. We do this because we use the dataset not only for the evaluation of our method for context-driven cross-document coreference resolution but also for the evaluation of a second approach, which aims to identify how the persons are portrayed (see Sect. 5.2). Section 4.3.3 introduces and describes our approach for context-driven cross-document coreference resolution. We then evaluate the approach (Sect. 4.3.4) and derive future work ideas (Sect. 4.3.5). Lastly, we summarize our research and set the results in context of the overall approach (Sect. 4.3.6).

In this section, for improved readability, we use the term *sentence-level bias* forms to refer to the three bias forms that cover the broad spectrum of text means to slant coverage, i.e., word choice and labeling, source selection, and commission

---

[5] To reflect recent developments in the field compared to the publication ([130]) summarized in this section, the following changes are made to this section by adapting parts of our recent paper on cross-document coreference resolution [403]. The discussion of related work is updated to include the latest literature (Sect. 4.3.1). The description of our preprocessing is updated to contain an improved variant that addresses an issue caused by CoreNLP (Sect. 4.3.3.1). The evaluation is updated to compare our method with the state of the art (Sect. 4.3.4).

and omission of information. We do not use the term person-targeting bias forms to highlight that our method can resolve various concept types and not only individual persons.

NewsWCL50 and its codebook are available at https://github.com/fhamborg/NewsWCL50.

4.3.1   Related Work

The task of coreference resolution entails techniques that aim to resolve mentions of entities, typically in a single text document. Coreference resolution is employed as an essential analysis component in a broad spectrum of use cases, such as identifying potential targets in sentiment analysis or as a part of discourse inter-pretation. While traditional coreference resolution focuses on single documents, *cross-document coreference resolution* (CDCR) resolves concept mentions across a set of documents. Compared to traditional coreference resolution, CDCR is a less-researched task. Moreover, CDCR can be considered more difficult than traditional coreference resolution since multiple documents yield a larger search space than only a single document. Adding to the difficulty, multiple documents are more likely to differ in their writing style (cf. "word choice" as described in Sect. 2.3.4). In this thesis, especially the varying word choice represents an important issue that current methods for coreference resolution and CDCR fail to tackle.

Only a few methods and datasets for CDCR have been proposed, especially compared to traditional single-document coreference resolution. Albeit evaluated on different datasets, the mildly decreased performance of CDCR can serve as an indicator for the increased difficulty and decreased research popularity ($F1 = [71.2; 79.5]$ [18]) compared to single-document coreference resolution ($F1 = 80.2$ [389]). Additionally, in initial experiments, we noticed strong performance losses when applying most techniques in a more realistic setup to reflect real-world use (cf. [403]). The key difference between established evaluation practices and real-world use is that no gold standard mentions are available in the latter. Instead, other techniques must first find and extract mentions before coreference resolution can resolve them. Naturally, such automated extraction is prone to errors, and imperfectly resolved concepts and mentions may degrade the performance of coreference resolution. To our knowledge, there is only one approach that jointly extracts and resolves mentions [200].

Missing the highly context-specific coreferences of varying word choice as they occur in biased news coverage is the fundamental shortcoming of prior CDCR. Identifying and resolving such mentions is especially important in person-oriented framing analysis (PFA). A fitting CDCR method would need to identify and resolve not only clearly defined concepts and identity coreferences. Additionally, the method would need to resolve near-identity mentions, such as in specific contexts "the White House" and "the US," and highly event-dependent coreferences, such as "Kim Jong-un" and "little rocket man" [74]. However, prior CDCR focuses on clearly defined concepts that are either event-centric or entity-centric. This narrowly defined structural distinction leads to corresponding methods and dataset annotations missing the previously mentioned concept types and highly context-dependent coreferences.

As stated previously, established CDCR datasets are either event-centric or entity-centric. When comparing the relevant datasets, we find a broad spectrum of *concept scopes*, e.g., whether two mentions are considered coreferential and which phrases are to be annotated as mentions in the first place. Correspondingly, individual datasets "miss" concepts and mentions that would have been annotated if the other annotation scheme had been used. The EventCorefBank (ECB) dataset entails two types of concepts, i.e., action and entity [23]. ECB+ is an event-centric corpus that extends ECB to consist of 502 news articles. Compared to ECB, annotations in ECB+ are more detailed, e.g., the dataset distinguishes various sub-types of actions and entities [62]. ECB+ contains only those mentions that describe an event, i.e., location, time, and human or non-human participants. NP4E is a dataset for entity-only CDCR [143]. NiDENT is an explorative CDCR evaluation dataset based on NP4E. Compared to the previously mentioned datasets, NiDENT also contains more abstract and less obvious coreference relations coined near-identity [299]. Zhukova et al. [403] provide an in-depth discussion of these and further datasets.

To our knowledge, all CDCR methods focus on resolving only events and, if at all, resolve entities as subordinate attributes of the events [174, 217]. There are two common, supervised approaches for event-centric CDCR: easy-first and mention-pair [217]. Easy-first models are so-called sieve-based models, where sieves are executed sequentially. Thereby, each sieve merges, i.e., resolves, mentions concerning specific characteristics. Initial sieves address reliable and straightfor-ward properties, such as heads of phrases. Later sieves address more complex or specialized cases using techniques such as pairwise scoring of pre-identified concepts with binary classifiers [158, 199, 216]. Recently, a mention-pair model was proposed, which uses a neural model trained to score the likelihood of a pair of events or entity mentions to be the same semantic concept. The model represents such mentions using spans of text, contexts, and semantic dependencies [18].

In sum, the reviewed CDCR methods suffer from at least one of three essential shortcomings. First, they only resolve clearly defined identity mentions. Second, they only focus on event-related mentions. Third, they suffer performance losses when evaluated in real-world use cases due to requiring gold standard mentions, which are not available in real-world use cases. These shortcomings hold corre-spondingly for the current CDCR datasets. Thus, to our knowledge, there is no CDCR method that resembles the annotation of persons and other concept types as established in framing analyses, including broadly defined concepts and generally concepts independent of fine-grained event occurrences. In the remainder of this section, we thus create a dataset and method that addresses these shortcomings.

4.3.2   NewsWCL50: Dataset Creation

To create *NewsWCL50*, the first dataset for the evaluation of methods for context-driven cross-document coreference resolution and methods to automatically identify sentence-level bias forms, we conducted a manual content analysis. Thereby, we follow the procedure established in the social sciences, e.g., we first use an inductive to explore the data and derive categories to be annotated as well as annotation instructions. Afterward, we conduct a deductive content analysis, following these instructions and using only these categories. NewsWCL50 consists of 50 news articles that cover 10 political news events, each reported on by 5 online US news outlets representing the ideological spectrum, e.g., including left-wing, center, and right-wing outlets. The dataset contains 8656 manual annotations, i.e., each news article has on average approximately 170 annotations.

4.3.2.1   Collection of News Articles

We selected ten political events that happened during April 2018 and manually collected for each event five articles. To ease the identification and annotation of sentence-level bias forms, we aimed to increase the diversity of both writing style and content. Therefore, we selected articles published by different news outlets and selected events associated with different topical categories. We selected five large, online US news outlets representing the political and ideological spectrum of the US media landscape [245, 364]:

- HuffPost (formerly The Huffington Post, far left, abbreviated *LL*)
- The New York Times (left, *L*)
- USA Today (center or middle, *M*)
- Fox News Channel (right, *R*)
- Breitbart News Network (far right, *RR*)

News outlets with different slants likely use different terms when reporting on the same topic (see Sect. 2.3.4), e.g., the negatively slanted term "illegal aliens" is used by RR, whereas "undocumented immigrants" is rather used by L when referring to DACA recipients.

To increase the content diversity, we aimed to gather events for each of the following political categories (cf. [95]): economic policy (focusing on US economy), finance policy, foreign politics (events in which the USA is directly involved), other national politics, and global interventions (globally important events, which are part of the public, political discourse).

Table 4.9 shows the collected events of NewsWCL50. One frequent issue during data gathering was that even major events were not reported on by all five news outlets; especially the far-left or far-right outlets did not report on otherwise popular events (which may contribute to a different form of bias, named *event selection*; see

Table 4.9 Overview of the events part of NewsWCL50. All dates are in 2018

ID	Date	Category Name	# Annotations
0	04/18	for Pompeo's meeting in PRK	684
1	04/19	nat Comey memos	711
2	04/20	glo PRK suspends nuclear tests	720
3	04/20	nat DNC sues Russia, Trump campaign	1153
4	04/24	for Macron and Trump meeting	1064
5	04/26	for Planning Trump's visit to the UK	621
6	04/29	nat Migrant caravan crosses into the US	938
7	04/30	nat Delays of US metal tariffs	784
8	04/30	eco Mueller's questions for Trump	881
9	04/30	glo Iran nucelar files	720

Sect. 2.3.1). We could not find any finance policy event in April that all five outlets reported on; hence, we discarded this category.

4.3.2.2   Training Phase: Creation of the Codebook

We create and use NewsWCL50 to evaluate two methods. The first method is for context-driven cross-document coreference resolution as used in target concept analysis. Additionally, we use NewsWCL50 to evaluate a method we will propose in Sect. 5.2, which aims to identify how persons are portrayed in news articles. This second method is used in the frame analysis component. Integrating the annotation of both properties, i.e., coreferential relations of mentions and how the mentions, e.g., of persons, are portrayed, approximates the manual procedure of frame analysis. Thus, in the following, we describe the creation of NewsWCL50, including coreferential mentions of persons (and other semantic concept types) and the framing categories representing how the persons are portrayed.

The goal of the training phase was to get an understanding of news articles concerning their types of mentions as well as the mentions' coreference relations and portrayal. The training phase was conducted on news articles not contained in NewsWCL50. We collected the articles as described in Sect. 4.3.2.1 but for different time frames. In a first, inductive content analysis, we asked three coders (students in computer science or political sciences aged between 20 and 29) to read five news articles and use MAXQDA, a content analysis software, to annotate any phrase that they felt was influencing their perception or judgment of a person and other semantic concept mentioned in the article. Specifically, coders were asked to (1) mark such phrases and state which (2) perception, judgment, or feeling the phrase caused in them, e.g., affection, and its (3) *target concept*, i.e., which concept the perception effect was ascribed to. We then used the initial codings to derive coding rules and a set of frame properties, representing how the annotators felt a target was portrayed.

We use the stated perception or judgment (see step 2 described previously) to derive so-called frame properties. Frame properties are pre-defined categories representing specific dimensions of political frames. A detailed definition and discussion is described in Sect. 5.2, where we introduce our method that determines the frame properties. Our desired characteristics of frame properties are on the one hand to be *general* so that they can be applied meaningfully to a variety of political news events, but on the other hand to be *specific*, allowing fine-grained categorization of persons' portrayal. Thus, during training, we added, removed, refined, or merged frame properties, e.g., we found that "unfairness" was always accompanied by (not necessarily physical) aggression and hence was better, i.e., more fine-grained, represented by "aggressor" or "victim," which convey additional information on the perception of the target. We created a codebook including definitions of frame properties, coding rules, and examples.

During training, we refined the codebook until we reached an acceptable inter-coder reliability (ICR) after six training iterations. The inter-coder reliability at the end of our training was 0.65 for frame properties and 0.86 for target mentions (calculated with mean pairwise average agreement). The comparably low inter-coder reliability of the annotations concerning frame properties is in line with results of other studies that aim to annotate topic-independent "frame types" [45]. This indicates the complexity and difficulty of the task.

In total, we derived 13 bi-polar frame properties, i.e., that have an antonym, and 3 without an antonym. Since the frame properties are not used in the target concept analysis, further details concerning the frame properties are described in Sect. 5.2.2, which also describes the method that uses the frame properties.

Target concepts can be "actors" (single individua), "actions," "countries," "events," "groups" (of individua acting collectively, e.g., demonstrators), other (physical) "objects," and also more abstract or broadly defined semantic concepts, such as "Immigration issues," coined "misc" (see Table 4.12). To define these seven types, we used established named entity types [359] and refined them during our annotation training to better fit our use case, e.g., by removing types that were never subject to bias, such as "TIME," and adding fine-grained sub-types, such as "countries" and "groups" instead of only "ORG."

The codebook is available as part of NewsWCL50 (see Sect. 4.3.6).

4.3.2.3 Deductive Content Analysis

To create our NewsWCL50 dataset, we conducted a deductive content analysis. One coder read and annotated the news articles, and two researchers reviewed and revised the annotations to ensure adherence to the codebook (cf. [260, 316]). For the annotation process, we used the two coding concepts devised in Sect. 4.3.2.2: *target concepts*, which are semantic concepts including persons that can be the target of sentence-level bias forms, and *frame properties*, which are categorized framing effects.

To facilitate the use of our dataset and method for context-driven cross-document coreference resolution also outside the scope of this thesis, we do not restrict the annotation of only persons but include the previously mentioned in total seven types of semantic concepts. This way, the dataset and the context-driven coreference resolution method we evaluate on it can more realistically cover the broad spectrum of coreferential mentions as they occur in real-world news coverage.

Following the codebook, the coder was asked to code any relevant phrase that represents either a target concept or frame property. This is in contrast to the beginning of the training phase, where any annotation originated from a change in perception or judgment of a concept. Said differently, while in training annotation the mentions of persons and other semantic concepts were only annotated as such if the coder felt the sentence or generally an (adjacent) expression changed the perception of such semantic concept, in the deductive annotation, we annotated all mentions of any semantic concept. To improve annotation efficiency, we asked the coder, however, to first briefly read the given news article and determine which semantic concepts are mentioned at least three times. Then, mentions of only these and semantic concepts that were identified in previously annotated news articles had to be annotated.

For each frame property, additionally, the corresponding target concept had to be assigned. For example, in "Russia seizes Ukrainian naval ships," "Russia" would be coded as a target concept (type "country"), and "seizes" as a frame property (type "Aggressor") with "Russia" being its target. Each mention of a target concept in a text segment can be targeted by multiple frame property phrases. More details on the coding instructions can be found in NewsWCL50's codebook. The dataset consists of 5926 target concept codings and 2730 frame property codings. NewsWCL50 is openly available in an online repository (see Sect. 4.3.6).

4.3.3   Method

Given a set of news articles reporting on the same event, our method finds and resolves mentions referring to the same semantic concepts. The method consists of the preprocessing, candidate extraction, and candidate merging as shown in Fig. 3.1. Our evaluation dataset (Sect. 4.3.2.3) is not sufficiently large to train a method that uses deep learning techniques, and a large-scale dataset would cause high cost, e.g., the OntoNotes dataset, commonly used to train methods for coreference resolution, consists of more than 2000 documents and more than 25000 coreference chains [7]. By devising a rule-based method as described in the following, we are able to reduce the otherwise high annotation cost (see also Sect. 4.3.2).

While the previous section reporting on the dataset creation (Sect. 4.3.2) described also information important for the frame analysis component, e.g., how we devised and annotated frame properties, this section describes only our method for context-driven cross-document coreference resolution. The method for the identification of frame properties as devised for the frame analysis components is described in Sect. 5.2.

In the target concept analysis, the goal of the first sub-task, *candidate extraction*, is to identify phrases that contain a semantic concept, i.e., phrases that could be the target of sentence-level bias forms (Sect. 4.3.3.1). We identify noun phrases (NPs) and verb phrases (VPs). We coin such phrases *candidate phrases*, and they compare to the mentions of target concepts annotated in the content analysis (Sect. 4.3.2.3). The goal of the second sub-task, *candidate merging*, is to merge candidates referring to the same semantic concept, i.e., groups of phrases that are coreferential (see Sect. 4.3.3.2). Candidate merging includes state-of-the-art coreference resolution, but also aims to find coreferences across documents and in a broader sense (see Sects. 4.3 and 4.3.1), e.g., "undocumented immigrants" and "illegal aliens."

4.3.3.1   Preprocessing and Candidate Extraction

We perform natural language preprocessing, including part-of-speech (POS) tagging, dependency parsing, full parsing, named entity recognition (NER), and coreference resolution [56, 57], using Stanford CoreNLP with neural models where available, otherwise using the defaults for the English language [224].

As initial *candidates*, we extract coreference chains, noun phrases (NPs), and verb phrases (VPs). First, we extract each coreference chain including all its mentions found by coreference resolution as a single candidate. Conceptually, this can be seen as an initial merging of candidates, since we merge all mentions of the coreference chain into one candidate. Second, we extract each NP found by the parser as a single candidate. We avoid long phrases by discarding any NP consisting of 20 or more words. If phrase contains one or more child NPs, we extract only the parent, i.e., longest, phrase. We follow the same extraction procedure for VPs. In the following, when referring to NPs, we always refer to VPs as well, if not noted otherwise.

We set a *representative phrase* for each candidate, which represents the candidate's meaning. For coreference chain candidates, we take the representative mention defined by CoreNLP's coreference resolution [334]. For NP-based candidates, we take the whole NP as the representative phrase. We use the representative phrases as one property to determine the similarity of candidates.

We also determine a candidate's *type*, which is one of the types shown in Table 4.10. For each phrase in a candidate, we determine whether its head is a "person," "group," or "country," using the lexicographer dictionaries from WordNet [239] and NE types from NER [88], e.g., "crowd" or "hospital," are of type "group." In linguistics, the head is defined as the word that determines a phrase's syntactic category [240], e.g., the noun "aliens" is the head of "illegal aliens," determining that the phrase is an NP. In WordNet, individual words, i.e., here the head words, can have multiple senses, e.g., "hospital" could be a building and an organization. We use WordNet's ranked list of senses for each head word, to determine the head word's most likely type. Specifically, for h, a head's sense s of rank $n_s = \{1, 2, 3, ...\}$, we define $m(s) = 1/n_s$ as a weighting factor. We then calculate the head's type score for each type t individually as follows:

Table 4.10 Candidate types identified during preprocessing

Candidate type	Definition	Example
person-ne	Single person (NE)	Biden
person-nes	Multiple persons (NE)	Democrats
person-nn	Single person (non-NE)	immigrant
person-nns	Multiple persons (non-NE)	officials
group-ne	Organization (NE)	Congress
group	Group of people, place (non-NE)	crowd, hospital
country-ne	Country (NE)	Germany
country	Location (non-NE)	country
misc	Abstract concepts	program

$$s(h, t) = \sum_{s \in W(h)} m(s) T(s, t), \tag{4.3}$$

where $W(h)$ yields all senses of h in WordNet and $T(s, t)$ returns 1 if the queried type t is identical to the sense s's type defined in WordNet, else 0. For a candidate c consisting of $h \in c$ head words, we then calculate c's type score for each type t individually as follows:

$$S(c, t) = \sum_{h \in c} s(h, t). \tag{4.4}$$

Lastly, we assign the type t to candidate c where $S(c, t)$ is maximized. This way, our fine-grained type determination well reflects the different senses a word can have and their likeliness.

If the candidate contains at least one NE mention, we set the NE flag. For example, if most phrases of a candidate are NE mentions of a "person," we set the candidate type "person-ne." If the type is a person, we distinguish between singular and plural by counting the heads' POS types: NN and NNP for singular and NNS and NNPS for plural. If a candidate is neither a "person," "group," nor "country," e.g., because the candidate is an abstract concept, such as "program," we set its type to "misc." We use the candidate types to determine which candidates can be subject to merging and for type-to-type-specific merge thresholds.

We refer to the previously described preprocessing as our *standard prepro-cessing*. Since CoreNLP is prone to merge (large) coreferential chains incorrectly [57], we propose a second preprocessing variant. Our *split preprocessing* executes an additional task after all tasks of standard preprocessing. It takes CoreNLP's coreference chains and split likely incorrectly merged chains and mentions into separate chains. To determine which mentions of a chain are likely not truly part of that chain, split preprocessing employs named entity linking. Specifically, it attempts to link each mention of a coreference chain to its Wikipedia page [112]. Given a coreference chain and its mentions, our preprocessing removes mentions having a different linked entity than the entity linked by the majority of the chain's mentions. For the removed mentions, our preprocessing creates new chains [242].

4.3.3.2 Candidate Merging

The goal of the sub-task candidate merging is to find and merge candidates that refer to the same semantic concept. Current methods for coreference resolution (see Sect. 4.3.1) cannot resolve abstract and broadly defined coreferences as they occur in sentence-level bias forms, especially in bias by word choice and labeling (see Table 1.1). Thus, we propose a merging method consisting of six sieves for our rule-based merging system (see Fig. 3.1), where each sieve analyzes specific characteristics of two candidates to determine whether the candidates should be merged.[6] Merging sieves 1 and 2 determine the similarity of two candidates, particularly of the "actor" type, as to their (core) meaning. Sieves 3 and 4 focus on multi-word expressions. Sieves 5 and 6 focus on frequently occurring words common in two candidates.

Sieves and Examples

(1) Representative phrases' heads: we merge two candidates by determining the similarity of their core meaning (as a simplified example, we would merge "Donald Trump" and "President Trump"). *(2) Sets of phrases' heads:* we determine the similarity as to the meaning of all phrases of two candidates ({Trump, president} and {billionaire}). *(3) Representative labeling phrases:* similarity of adjectival labeling phrases. Labeling is an essential property in sentence-level bias forms, especially in bias by word choice and labeling ("illegal immigrants" and "undocumented workers"). *(4) Compounds:* similarity of nouns bearing additional meaning to the heads ("DACA recipients" and "DACA applicants"). *(5) Representative wordsets:* similarity of frequently occurring words common in two candidates ("United States" and "US"). *(6) Representative frequent phrases:* similarity of longer multi-word expressions where the order is important for the meaning ("Deferred Action of Childhood Arrival" and "Childhood Arrivals").

For each merging sieve i, we define a 9×9 comparison matrix cmat_i spanned over the nine candidate types listed in Table 4.3. The normalized scalar in each cell $\text{cmat}_{i,u,v}$ defines whether two candidates of types u and v are considered comparable (if $\text{cmat}_{i,u,v} > 0$). As described later, for some merging sieves, we also use $\text{cmat}_{i,u,v}$ as a threshold, i.e., we merge two candidates with types u and v if the similarity of both candidates is larger than $\text{cmat}_{i,u,v}$. We found generally usable default values for the comparison matrices' cells and other parameters described in the following through experimenting and domain knowledge (see Sect. 4.3.5). The specific values of all comparison matrices can be found in the source code (Appendix A.4).

---

[6] For a visual depiction of the sieve-based process, please refer to Appendix A.4.

We organize candidates in a list sorted by their number of phrases, i.e., mentions in the texts; thus, larger candidates are at the top of the list. In each merging sieve, we compare the first, thus largest, candidate with the second candidate, then third, etc. If two candidates at comparison meet a specific similarity criterion, we merge the current (smaller) candidate into the first candidate, thereby removing the smaller candidate from the list. Once the pairwise comparison reaches the end of the list for the first candidate, we repeat the procedure for each remaining candidate in the list, e.g., we compare the second (then third, etc.) candidate pairwise with the remainder of the list. Once all candidates have been compared with another, we proceed with the next merging sieve.

As stated previously, the first and second sieve aim to determine the similarity of two candidates as to their (core) meaning. In the **first merging sieve**, we merge two candidates if the *head* of each of their *representative phrase* (see Sect. 4.3.3.1) is identical by string comparison. By default, we apply the first merging sieve only to candidates of identical NE-based types, but one can configure the sieve's comparison table $cmat_1$ to be less restrictive, e.g., allow also other type comparison or inter-type comparisons.

In the **second merging sieve**, we merge two candidates if their *sets of phrases' heads* are semantically similar. For each candidate, we create a set H consisting of the heads from all phrases belonging to the candidate. We then vectorize each head within H into the word embedding space of the enhanced word2vec model trained on the GoogleNews corpus (300M words, 300 dimensions) [238], using an implementation that also handles out-of-vocabulary words [280]. We then compute the mean vector $\overrightarrow{m_H}$ for the whole set of head vectors.

Then, to determine whether two candidates c_0 and c_1 are semantically similar, we compute their similarity $s(c_0, c_1) = \text{cossim}(\overrightarrow{m_H}, \overrightarrow{n_H})$, where $\overrightarrow{m_H}$ is the mean head vector of c_0, $\overrightarrow{n_H}$ the mean head vector of c_1, and cossim(...) the cosine similarity function. We merge the candidates, if c_0 and c_1 are of the same type, e.g., each represents a person, and if their cosine similarity $s(c_0, c_1) \geq t_{2,\text{low}} = 0.5$. We also merge candidates that are of different types if we consider them comparable (defined in $cmat_2$), e.g., NEs such as "Trump" with proper nouns (NNP) such as "President," and if $s(c_0, c_1) \geq t_{2,\text{high}} = 0.7$. We use a higher, i.e., more restrictive, threshold since the candidates are not of the same type.

The third and fourth sieves focus on resolving multi-word expressions, such as "illegal immigrants" and "undocumented workers." In the **third merging sieve**, we merge two candidates if their *representative labeling phrases* are semantically similar. First, we extract all *adjective NPs* from a candidate containing a noun and one or more labels, i.e., adjectives attributing to the noun. If the NP contains multiple labels, we extract for each label one NP, e.g., "young illegal immigrant" is extracted as "young immigrant" and "illegal immigrant." Then, we vectorize all NPs of a candidate and cluster them using affinity propagation [91]. To vectorize each NP, we concatenate its words, e.g., "illegal_worker," and look it up in the embedding space produced by the enhanced word2vec model (see second merging sieve), where frequently occurring phrases were treated as separate words during training [197].

If the concatenated NP is not part of the model, we calculate a mean vector of the vectors of the NP's words. Each resulting cluster consists of NPs that are similar in meaning. For each cluster within one candidate, we select the single adjective NP with the global most frequent label, i.e., the label that is most frequent among all candidates. This way, selected NPs are the *representative labeling phrases* of a candidate.

Then, to determine the similarity between two candidates c_0 and c_1 in the third merging sieve, we compute a similarity score matrix $S(V, W)$ spanned by the representative labeling phrases $v_i \in V$ of c_0 and $w_j \in W$ of c_1. We look up a type-to-type-specific threshold $t_3 = \text{cmat}_3 \left[\text{type}(c_0) \right] [\text{type}(c_1)]$, and type (c) returns the type of candidate c (see Table 4.3). For each cell $s_{i,j}$ in $S(V, W)$, we define a three-class similarity score

$$s_{i,j} = \begin{cases} 2, & \text{if } \text{cossim}(\overrightarrow{v_i}, \overrightarrow{w_j}) \geq t_3 + t_{3,r} \\ 1, & \text{if } \text{cossim}(\overrightarrow{v_i}, \overrightarrow{w_j}) \geq t_3 \\ 0, & \text{otherwise,} \end{cases} \qquad (4.5)$$

where $\text{cossim}\left(\overrightarrow{v_i}, \overrightarrow{w_j}\right)$ is the cosine similarity of both vectors and $t_{3,r} = 0.2$ to reward more similar vectors into the highest similarity class. We found the three-class score to yield better results than using the cosine similarity directly. We merge c_0 and c_1 if $V \sim W$, i.e., $\text{sim}(V, W) = \frac{\sum_{s \in S} s_{i,j}}{|V||W|} \geq t_{3,m} = 0.3$. When merging candidates, we transitively merge different candidates U, V, W if $U \sim W$ and $V \sim W$, i.e., we say $U \sim W, V \sim W \rightarrow U \sim V$, and merge both candidates U and W into V.

In the **fourth merging sieve**, we merge two candidates if they contain *compounds* that are semantically similar. In linguistics, a compound is a word or multi-word expression that consists of more than one stem and that cannot be separated without changing its meaning [209]. We focus only on multi-word compounds, such as "DACA recipient."

In this sieve, we first analyze the semantic similarity of the stems common in multiple candidates. Therefore, we find all words that are common in at least one compound of each candidate at comparison. In each candidate, we then select as its *compound phrases* all phrases that contain at least one of these words and vectorize the compound phrases into the word embedding space. Then, to determine the similarity of two candidates, we compute a similarity score matrix $S(V, W)$ spanned by all compound phrases $v_i \in V$ of candidate c_0 and $w_j \in W$ of c_1 using the same approach we used for the third merging sieve (including merging candidates that are transitively similar). If $\text{sim}(V, W) \geq t_{4,m}$, we merge both candidates. Else, we proceed with the second merge method.

In the second method, we check for the lexical identity of specific stems in multiple candidates. Specifically, we merge two candidates c_0 and c_1 if there is at least one phrase in c_0 that contains a head that is a dependent in at least one phrase in c_1 and if both candidates are comparable according to cmat_4. For instance, two candidates are of type person-ne (see Table 4.3), and one phrase in c_0 has a headword "Donald," and one phrase in c_1 is "Donald Trump," where "Donald" is the dependent word.

The fifth and sixth sieves focus on the special cases of coreferences that are still unresolved, particularly candidates that have frequently occurring words in common, such as "United States" and "US." In the **fifth merging sieve**, we merge two candidates if their *representative wordsets* are semantically similar. To create the representative wordset of a candidate, we perform the following steps. We create frequent itemsets of the words contained in the candidate's phrases excluding stopwords (we currently use an absolute support supp = 4) and select all maximal frequent itemsets [3]. Note that this merging sieve thus ignores the order of the words within the phrases. To select the most representative wordsets from the maximal frequent itemsets, we introduce a representativeness score

$$r(w) = \log(1 + l(w))\log(f(w)), \tag{4.6}$$

where w is the current itemset, $l(w)$ the number of words in the itemset, and $f(w)$ the frequency of the itemset in the current candidate. The representativeness score balances two factors: first, the *descriptiveness* of an itemset, i.e., the more words an itemset contains, the more comprehensively it describes its meaning, and second, the *importance*, i.e., the more often the itemset occurs in phrases of the candidate, the more relevant the itemset is. We then select as the *representative wordsets* the N itemsets with the highest representativeness score, where $N = \min(6, f_p(c))$, where $f_p(c)$ is the number of phrases in a candidate. If a word appears in more than $rs_5 = 0.9$ of all phrases in a candidate but is not present in the maximal frequent itemsets, we select only $N - 1$ representative wordsets and add an itemset consisting only of that word to the representative wordsets. Lastly, we compute the mean vector $\vec{v}$ of each representative wordset v by vectorizing each word in the representative wordset using the word embedding model introduced in the second merging sieve.

Then, to determine the similarity of two candidates c_0 and c_1 in the fifth merging sieve, we compute a similarity score matrix $S(V, W)$ spanned by all representative wordsets $v_i \in V$ of candidate c_0 and $w_j \in W$ of c_1 analogously constructed as the matrix described in the third merging sieve. We merge c_0 and c_1, if sim $(V, W) \geq t_5 = 0.3$.

In the **sixth merging sieve**, we merge two candidates if they have similar *representative frequent phrases*. To determine the most representative wordlists of a candidate, we conceptually follow the procedure from the fifth merging sieve but apply the steps to phrases instead of wordsets. Specifically, the representativeness score of a phrase o is calculated using Eq. 4.6 with phrase o instead of itemset w. We then select as the *representative frequent phrases* the N phrases with the highest representative score, where $N = \min 6, f_p(c)$.

Then, to determine the similarity of two candidates c_0 and c_1 in the sixth merging sieve, we compute a similarity score matrix $S(V, W)$ spanned by all representative wordlists $v_i \in V$ of candidate c_0 and $w_j \in W$ of c_1. We look up a type-to-type-specific threshold $t_6 = \text{cmat}_6 \left[\text{type}(c_0)\right] [\text{type}(c_1)]$. We calculate the similarity score of each cell $s_{i,j}$ in $S(V, W)$:

$$s_{i,j} = \begin{cases} 2, & \text{if levend}\left(v_i,\ w_j\right) \leq t_6 - t_{6,r} \\ 1, & \text{if levend}\left(v_i,\ w_j\right) \leq t_6 \\ 0, & \text{otherwise,} \end{cases} \tag{4.7}$$

where $\text{levend}\left(v_i,\ w_j\right)$ is the normalized Levenshtein distance [204, 226] of both phrases and $t_{6,r} = 0.2$. Then, over all rows j, we find the maximum sum of similarity scores sim_{hor} and likewise sim_{vert} over all columns i:

$$\text{sim}_{\text{hor}} = \max_{0 \leq i < |W|} \left(\sum_{j=0}^{|V|} s_{i,j} \right) / |W| \tag{4.8}$$

and

$$\text{sim}_{\text{vert}} = \max_{0 \leq j < |V|} \left(\sum_{i=0}^{|W|} s_{i,j} \right) / |V| . \tag{4.9}$$

We calculate a similarity score for the matrix:

$$\text{simval}(V, W) = \begin{cases} \text{sim}_{\text{hor}}, & \text{if } \text{sim}_{\text{hor}} \geq \text{sim}_{\text{vert}} \land |W| > 1 \\ \text{sim}_{\text{vert}}, & \text{else if } |V| > 1 \\ 0, & \text{otherwise.} \end{cases} \tag{4.10}$$

Finally, we merge candidates c_0 and c_1 if $\text{simval} \geq t_{6,m} = 0.5$.

Using the previously outlined series of six sieves, our method merges those candidates that CoreNLP's coreference resolution used in the preprocessing step did not identify as coreferential. In practical terms, our method relies on CoreNLP as an established method for single-document coreference resolution and uses the six sieves to enhance CoreNLP's results in two ways. First, our method merges the coreferences found in single documents across multiple documents. Second, it additionally merges highly context-dependent coreferences as they occur frequently in the sentence-level bias forms that our PFA approach seeks to identify.

4.3.4   Evaluation

To measure the effectiveness of the context-driven cross-document coreference resolution in the context of our overall analysis for the automated identification and communication of media bias, we perform an in-depth evaluation of the approach in this section. After presenting the evaluation results in Sect. 4.3.4.3, we discuss the strengths and weaknesses of our approach, from which we derive future research directives in Sect. 4.3.5.

4.3.4.1 Setup and Metrics

We evaluate our method and all baselines on the events and articles contained in the NewsWCL50 dataset (Sect. 4.3.2.3). Similar to prior work, we evaluate exclusively the coreference resolution performance but not the extraction performance [18]. Thus, we do not automatically extract the mentions for coreference resolution but pass the set of all true mentions as annotated in NewsWCL50 to evaluate the method.

Our primary evaluation metric is weighted macro F1 ($F1_m$), where we weight the F1 score of each candidate (as automatically resolved) by the number of samples from its true class (as annotated in our dataset). Secondary metrics are precision (P) and recall (R). We generally prefer recall over precision within the secondary metrics since CoreNLP is prone to yield many small or even singleton coreference chains on our dataset. CoreNLP thus achieves very high precision scores, while at the same time, the larger coreference chains miss many mentions, i.e., those mentions that are part of the singleton chains. To measure the metrics, we compare resolved candidates, i.e., coreference chains, with manually annotated target concepts, e.g., "USA/Donald Trump." For each target concept annotated in NewsWCL50, we find the best matching candidate extracted automatically (cf. [226]), i.e., the candidate whose phrases yield the largest overlap to the mentions of the target concept. To account for the subjectivity of the coding task in the content analysis, particularly when coding abstract target concepts (Sect. 4.3.2), we allow in our evaluation multiple true labels to be assigned to the candidates, i.e., a predicted candidate can have multiple true annotated target concepts.

We report and discuss all the performances of the evaluated methods for all concept types. However, we use the "Actor" type as the primary concept type since the person-oriented framing approach focuses on the analysis of individual persons (Sect. 3.3.2).

4.3.4.2 Baselines

We compare our approach with three baselines, which we describe briefly in the following.

EECDCR represents the state of the art in cross-document coreference resolution. The authors reported the highest evaluation results [18] compared to Kenyon-Dean, Cheung, and Precup [170], Lee et al. [199], and NLP Architect [158]. EECDCR resolves event and entity mentions jointly. To reproduce EECDCR's performance, we use the model's full set of optional features: semantic role labeling (SRL), ELMo word embeddings [284], and dependency relations. Since we could not setup

SwiRL, the originally used SRL parser, we used the most recent SRL method by AllenNLP [323]. To resolve intra-document mentions, we use CoreNLP [56]. We use default parameters for the model inference.

Two further baselines represent the state of the art in single-document coreference resolution; both employ CoreNLP [56, 57]. Since CoreNLP is designed for single-document coreference resolution, each baseline uses a different adaption to make CoreNLP suitable for the cross-document task. Baseline *CoreNLP-merge* creates a virtual document by appending all documents. It then performs CoreNLP's coreference resolution on this virtual document. Baseline *CoreNLP-cluster* employs CoreNLP's coreference resolution on each document individually. The baseline then clusters the mentions of all coreference chains in the word2vec space [197] using affinity propagation [91]. Each resulting cluster of phrases yields one candidate (cf. [43]), i.e., coreference chain.

4.3.4.3   Results

Table 4.11 shows the CDCR performance of the evaluated methods. Our method achieved the highest performance concerning our primary metric ($F1_m = 59.0$). The CoreNLP baselines yield worse performance (at best: $F1_m = 53.2$). Our method performed also slightly better than EECDCR ($F1_m = 57.8$). We found that our split preprocessing tackled CoreNLP's merging issue, where large chains are merged incorrectly as described in Sect. 4.3.3.1. Specifically, our split preprocessing improved the $F1_m$ performance from at best 57.0 to 59.0. The effect of our split preprocessing can be seen when comparing the recall scores after preprocessing

Table 4.11 Performance of the context-driven cross-document coreference resolution method and baselines on NewsWCL50. Best-performing approaches are highlighted

Method	Preprocessing	P	R	F1
CoreNLP-merge		90.7	34.8	43.1
CoreNLP-cluster		79.4	44.8	**53.2**
EECDCR		86.3	49.6	**57.8**
Preprocessing		95.5	15.8	25.6
Sieve 1		88.4	35.4	42.8
Sieves 1 to 2		81.1	51.1	56.8
Sieves 1 to 3	standard	80.9	51.2	56.8
Sieves 1 to 4		80.4	51.4	**57.0**
Sieves 1 to 5		75.1	**53.5**	56.3
Sieves 1 to 6		75.1	**53.5**	56.3
Preprocessing		95.0	34.2	44.2
Sieve 1		88.7	40.0	47.8
Sieves 1 to 2		84.3	53.2	**59.0**
Sieves 1 to 3	split	83.5	53.2	58.8
Sieves 1 to 4		81.6	53.5	58.8
Sieves 1 to 5		76.8	**56.1**	58.5
Sieves 1 to 6		76.8	**56.1**	58.5

Table 4.12 Performance of the context-driven cross-document coreference resolution method and baselines for each concept type. Macro F1 is shown for each of the three baselines: CoreNLP-merge, CoreNLP-cluster, and EECDCR

Concept type	Support	merge	cluster	EECDCR	P	R	F1
Action	124	8.0	37.9	**39.7**	53.4	37.9	**39.7**
Actor	1336	75.3	74.7	81.9	91.6	87.4	**88.7**
Actor-I	512	26.0	**37.3**	34.8	84.0	28.1	36.4
Country	1569	44.2	47.4	**53.1**	61.3	48.6	46.0
Event	493	36.1	55.6	**63.6**	62.2	67.1	60.8
Group	498	27.3	35.9	39.6	70.9	34.7	**43.1**
Misc	514	11.7	38.0	42.8	95.8	32.0	**43.2**
Object	583	38.1	61.8	64.2	83.7	67.1	**71.3**

(standard preprocessing, $R = 15.8$; split preprocessing, $R = 34.2$). Since our method performed better with split preprocessing, we refer in the remainder of this section to this variant if not noted otherwise.

Table 4.12 shows the performance achieved on the individual concept types. Importantly, our method achieved the highest performance on the "Actor" type ($F1_m = 88.7$ compared to the best baseline: $F1_m = 81.9$). The high "Actor" performance allows for using our method in target concept analysis, since person-oriented framing analysis focuses on individual persons, i.e., as part of the "Actor" type. Further, our method performed for most of the other concept types better than the baselines: "Action," "Group," "Misc," and "Object." On the "Event" type, EECDCR performed best, which is expected since the method was specifically designed for event-centric coreference resolution ($F1_m = 63.6$ compared to us: $F1_m = 60.8$).

In general, we found that our method and the baselines performed best on concepts that consist mainly of NPs and that are narrowly defined. In contrast, our method performed worse on concepts that consist mainly of *(1)* VPs, are *(2) broadly defined*, or are *(3) abstract*. Our method achieved a low macro F1 on the "Action" type ($F1_m = 39.7$), whose candidates consist mainly of VPs. The concept type "Actor-I" is very broadly defined as to our codebook and yields the lowest performance ($F1_m = 36.4$). One reason for the low performance is that in the content analysis, different individua were subsumed under one Actor-I concept to increase annotation speed (Sect. 4.3.2). We propose means to address this issue in Sect. 4.3.5. The extraction of candidates of the type "Misc" is as expected challenging ($F1_m = 43.2$), since by definition its concepts are mostly abstract or complex. For example, the concept "Reaction to IRN deal" (event #9) contains both actual and possible future (re)actions to the event (the "Iran deal") and assessments and other statements by persons regarding the event.

Table 4.13 shows the performance of our method on the individual events of NewsWCL50. The approach performed best on events #1, #4, and #9 ($F1_{m,1} = 68.2$) and worse on events #6, #7, and #3 ($F1_{m,6} = 46.2$). When investigating

Table 4.13 Performance of the context-driven cross-document coreference resolution method for each event

Event ID	Support	P	R	F1
0	548	66.4	63.8	59.1
1	612	90.9	63.0	**68.2**
2	608	74.3	54.8	59.6
3	835	79.8	45.4	52.2
4	628	77.6	67.2	64.7
5	444	70.9	60.8	60.0
6	583	75.5	39.8	**46.2**
7	577	63.8	53.0	50.9
8	641	81.9	58.6	62.1
9	498	84.4	60.3	64.7

the events' compositions of concept types, we found that in general higher performances were achieved for events that consist mainly of NPs, e.g., 44.1% of all mentions in event #1 are of type "Actor."[7] In contrast, events with lower performance contain typically a higher number of broadly defined concepts or "Action" concepts.

We also found that our approach was able to extract and merge unknown concepts, i.e., concepts that are not contained in the word embeddings used during the candidate merging process [197, 280]. For example, when the GoogleNews corpus was published in 2013 [238], many concepts, such as "US President Trump" or "Denuclearization," did not exist yet or had a different, typically more general, meaning than in 2018. Yet, the approach was able to correctly merge phrases with similar meanings, e.g., in event #2, the target concept "Peace" contains among others "a long-term detente," "denuclearization," and "peace." In event #6, the approach was able to resolve, for example, "many immigrants," "the caravan," "the group marching toward the border," "families," "refugees," "asylum seekers," and "unauthorized immigrants." In event #1, the approach resolved, among others, "allegations," "the infamous Steele dossier," "the salacious dossier," and "unsubstantiated allegations."

Table 4.11 shows that using only sieves 1 to 2 achieved over all concept types the highest performance ($F1_m = 59.0$). However, subsequent sieves improved recall further (from $R = 53.2$ after sieve 2 to $R = 56.1$ after sieve 6) while only slightly reducing $F1_m$ to 58.5. We thus recommend to generally run all sieves. However, in the context of person-oriented framing analysis (PFA), we recommend to use only sieves 1 and 2. Table 4.14 shows that using only these two sieves suffices to achieve the best possible performance for the "Actor" type ($F1_m = 88.8$). This is expected since sieves 1 and 2 focus specifically on resolving mentions of the "Actor" type.

In sum, the results of the evaluation showed an improved performance of our method in resolving highly context-dependent coreferences compared to the state of the art, especially on the *Actor* concept type, which is most relevant for PFA.

---

[7] An overview of the event composition can be found in Table A.2 in Sect. A.2.

Table 4.14 Performance of
the context-driven
cross-document coreference
resolution method evaluated
only on the "Actor" type

Method	P	R	F1
Preprocessing	97.7	72.4	81.2
Sieve 1	92.4	83.5	86.9
Sieves 1 to 2	91.6	87.4	**88.8**
Sieves 1 to 3	91.6	87.4	88.8
Sieves 1 to 4	91.6	87.4	88.8
Sieves 1 to 5	91.6	87.4	88.7
Sieves 1 to 6	91.6	87.4	88.7

4.3.5   Future Work

When devising our method, we focused on using it only as part of person-oriented framing analysis (PFA). In this use case, i.e., resolving individual persons (concept type: "Actor") in event coverage, our method outperformed the state of the art in coreference resolution ($F1_m = 88.7$ compared to 81.9). Moreover, the evaluation showed that our method achieved competitive or higher performance compared to current methods for cross-document coreference resolution when evaluated on all concept types or individual concept types. However, our evaluation cannot elucidate how effective our method is in other domains and applications. We seek to address this by creating a larger dataset, for which we plan to implement and validate minor improvements in the codebook, e.g., infrequent individua are currently coded jointly into a single "[Actor]-I" target concept. While such coding requires less coding effort, it also negatively skews the measured evaluation performance (see Sect. 4.3.4). An idea is to either not code infrequent target concepts or code them as single concepts.

To further strengthen the evaluation of the coreference resolution task, we plan to test our method on established datasets for coreference resolution (Sect. 4.3.1). Doing so will help to investigate how our method performs in other use cases and domains. Albeit standard in the evaluation practices of coreference resolution, the evaluation of methods only on true mentions means that the practical performance of such methods may differ strongly. We expect that the performance of a CDCR methods is lower when employed in real-world use cases where almost never true mentions, e.g., as annotated in a ground truth, are available. Instead, in real-world use cases, such mentions typically have to be extracted automatically. We thus propose to conduct a future evaluation that uses automatically extracted mentions, where we compare pure coreference resolution performance with its performance when employed in settings resembling real-world use.

A larger dataset would also enable the training of recent deep learning models, and we expect that deep learning models could achieve higher performance. Recent models for single-document coreference resolution achieve increased performance compared to earlier, traditional methods [390]. A key issue that prevented us from creating a sufficiently large dataset is the tremendous annotating cost required for its annotation (see Sect. 4.3).

4.3.6 Conclusion

The previous sections introduced our method for context-driven cross-document coreference resolution. Our method is the first to find and resolve coreferences as they are relevant for sentence-level bias forms. When evaluated on our dataset for the PFA use case, our method outperformed the state of the art (on the concept type "Actor," our method achieved $F1_m = 88.7$ compared to 81.9). When considering all concept types, our method performed slightly better than the state of the art ($F1_m = 59.0$ compared to 57.8).

As noted in Sect. 4.3.5, our use case-specific evaluation can only serve as a first indicator for the general coreference resolution performance in other use cases and on other text domains. Moreover, coreference resolution is a broad and complex field of research, with a diverse spectrum of sub-tasks and use cases. Our thesis can only contribute a first step in the task of context-driven cross-document coreference resolution. Despite that, because of the high performance of our method on the "Actor" type as evaluated on the NewsWCL50 dataset, we conclude that the method is an effective means to be used within the PFA approach.

NewsWCL50 and its codebook are available at https://github.com/fhamborg/NewsWCL50.

4.4 Summary of the Chapter

This chapter introduced *target concept analysis* as the first analysis component in person-oriented framing analysis (PFA). Target concept analysis aims to find and resolve the mentions of persons across the given news articles. This task is of particular importance and difficulty in slanted coverage, for example, due to the divergent word choice across differently slanted news articles. We explored two approaches to enable target concept analysis: event extraction and coreference resolution.

Our approach for *event extraction* (Sect. 4.2) achieved overall high performance in extracting answers for the journalistic 5W1H questions, which describe an article's main event. However, additional research effort would be required to employ this approach in target concept analysis. First, we would need to extend the approach to extract not only the single main event of each article but also multiple side events. The latter is crucial for PFA, which requires multiple mentions and persons to identify the overall frame of a news article. In contrast, a lower reliability of the frame identification would be expected when relying on only a single characteristic of each article, e.g., the actor of the single main event. Second, for the event extraction to be used in PFA, we would need to devise a second method for resolving the events and their phrases across the event coverage. Besides these two shortcomings concerning the use in PFA, our event extraction approach represents a universally usable means for event extraction, achieves high extraction performance, and is publicly available (Sect. 4.2.5).

Due to the two shortcomings of the line of research employing event extraction, we decided to focus on a second line of research. The main contribution of this chapter is the first method to perform *context-driven cross-document coreference resolution* (Sect. 4.3). In the evaluation, our method resolved highly context-dependent mentions of persons and other concept types as they occur commonly in person-targeting forms of media bias. When considering mentions of individual persons as relevant for PFA, our method achieved a high performance ($F1_m = 88.7$) and outperformed the state of the art ($F1_m = 81.9$). We will thus use our method for coreference resolution for the target concept analysis component.

Chapter 5
Frame Analysis

Abstract This chapter details the last component of person-oriented framing analysis: frame analysis. The component aims to classify how persons are portrayed in news articles. The chapter introduces and discusses two approaches for this task. First, it briefly presents an exploratory approach that aims to classify fine-grained categories of how persons are portrayed. Afterward, the chapter introduces the first method for target-dependent sentiment classification in the domain of news articles. The dataset and method enable sentiment classification in a domain that could not reliably be analyzed earlier. Lastly, the chapter argues for using the latter approach in the frame analysis component, in particular because of its high classification performance.

5.1 Introduction

This chapter describes the frame analysis component, which aims to identify how persons are portrayed in the given news articles, both at the article and sentence levels. This task is of difficulty for various reasons, such as that news articles rather implicitly or indirectly frame persons, for example, by describing actions performed by a person. Consequently, a high level of interpretation is necessary to identify how news articles portray persons. Because of this and other issues highlighted later in this chapter, prior approaches to analyze frames or derivatives yield inconclusive or superficial results or require high manual effort, e.g., to create large annotated datasets.

Frame analysis is the second and last analysis component in person-oriented framing analysis (PFA). The input to frame analysis is a set of news articles reporting on the same policy event, including the persons involved in the event and their mentions across all articles. The output of the frame analysis component should be information concerning how each person is portrayed and groups of articles similarly framing the persons involved in the event.

An approach in principle suitable for the frame analysis component is identifying political frames as defined by Entman [79]. Doing so would approximate the content analysis, particularly the frame analysis, as conducted in social science research on media bias most closely. However, as pointed out in Sect. 3.3.2, taking such an approach would result in infeasibly high annotation cost. Further, political frames are defined for a specific topic or analysis question [79], whereas PFA is meant to analyze bias in news coverage on any policy event. Thus, identifying political frames seems out of the thesis's scope and also methodologically infeasible currently. We revisit this design decision later in the chapter and also in our discussion of the thesis's limitations and future work ideas in Sect. 7.3.2.

In this chapter, we explore two conceptually different approaches to determine how individual persons are portrayed. The first approach more closely resembles how researchers in the social sciences analyze framing. It aims to identify categories representing topic-independent framing effects, which we call *frame properties*, such as whether a person is portrayed as being competent, wise, powerful, or trustworthy (or not). The second approach follows a more pragmatic route to the task of the frame analysis component, which we devise to address fundamental issues of the first approach, such as high annotation cost, high annotation difficulty, and low classification performance. Both approaches have in common that they do not analyze frames, which would be the standard procedure in the social sciences, but instead categorized effects of framing. We focus on framing effects since frames as analyzed in social science research on media bias are topic-specific. In contrast, our approach is meant to analyze news coverage on any policy issue (see also Sects. 1.3 and 3.3.2).

Note that the frame analysis component consists of an additional second task, i.e., frame clustering (Fig. 3.1). Once the frame analysis has identified how each person is portrayed at both the article and sentence levels, frame clustering aims to find groups of articles that frame the persons similarly. For frame clustering, we use a simple technique that we will describe when introducing our prototype system (Chap. 6).

5.2 Exploring Person-Targeting Framing(-Effects) in News Articles

The section presents the results of a research direction we pursued initially for the frame analysis component. We explore a simple approach that aims to identify so-called frame properties, which are fine-grained constituents of how a person is portrayed at the sentence level, e.g., whether a person is shown as competent or powerful. Frame properties resemble framing effects in social science research. We propose frame properties as pre-defined categorical characteristics that might be attributed to a target person due to one or more frames. For example, in a sentence that frames immigrants as intruders that might harm a country's culture and economy (rather than victims that need protection, cf. [369]), respective frame properties of the mentioned immigrants could be "dangerous" and "aggressive."

The remainder of this section is structured as follows. Section 5.2.1 briefly summarizes prior work on automated frame analysis. We then present our exploratory approach for frame property identification in Sect. 5.2.2. Afterward, we discuss the results of an exploratory, qualitative analysis in Sect. 5.2.3. Section 5.2.4 highlights the shortcomings and difficulties of this approach and discusses how to address or avoid them. Specifically, our approach achieved in the evaluation only mixed results, but we can use the identified issues to derive our main approach for the frame analysis component, which is then described in Sect. 5.3. Lastly, Sect. 5.2.5 provides a brief summary of our exploratory research on frame property identification.

The dataset used for the approach and its codebook are available at https://github.com/fhamborg/NewsWCL50.

5.2.1   Related Work

This section briefly summarizes key findings of our literature review concerning the analysis of political framing. An in-depth discussion of the following and other related approaches can be found in Chap. 2.

To analyze how persons (or other semantic concepts) are portrayed, i.e., framed, researchers from the social sciences primarily employ manual content analyses and frame analyses (see Sect. 2.3.4). In content analyses focusing on the person-targeting bias forms (Sect. 3.3.2), social scientists typically analyze how news articles frame individual persons or groups of persons. For example, whether there is a systematic tendency in coverage to portray immigrants in a certain way, such as being aggressive or helpless [263]. Observing these tendencies may then yield specific frames, such as the "intruder" or "victim" frames mentioned in the beginning of Sect. 5.2. Other analyses not concerned with persons but, e.g., topics, may focus on whether news outlets use emotional or factual language when reporting on a specific topic or which topical aspects of an issue are highlighted in coverage [274].

In sum, in our literature review on identifying media bias, we find that no automated system focuses on the analysis of person-targeting bias forms at the sentence level (see Sects. 2.3.2–2.3.4). However, two prior works are of special interest. First, the research conducted for the creation of the media frame corpus (MFC) aims to directly represent political frames [80] as established in the social sciences [45]. In contrast to political frames, MFC's "frame types" are topic independent and thus are in principle highly relevant for our task. However, from a conceptual perspective, the MFC's frame types are independent of any target, i.e., they holistically describe the content or "frame" of a news article or a sentence within. Moreover, this approach suffers from high annotation cost and low inter-coder reliability (ICR) [45]. As a consequence, classifiers trained on the MFC yield low classification accuracy on the sentence level [46].

Second, a recent approach aims to automatically extract so-called microframes from a set of text documents, e.g., news articles [193]. Given a set of text documents, the respective microframes are defined as semantic axes that are over-represented in individual documents. Each such semantic axis is a bi-polar adjective pair as used in semantic differential scales established in psychology [145]. Since the microframes are extracted for a given set of documents, they are topic-dependent or, more specifically, dataset-dependent. For example, in a topic and corresponding documents on immigration, the adjective pair of one microframe could be "illegal"-"legal." After the extraction of these microframes for a given dataset, users then review the microframes and select a subset of them to be used for further analysis. The approach yields qualitative microframes that resemble closely our frame properties but are, in contrast to them, dataset-specific.

Most of the other reviewed approaches that are only partially related to our task use quantitative methodology, and their results are mostly superficial, especially when compared to the results of manual content analyses. For example, one approach investigates the frequency of affective words close to user-defined words [116], e.g., names of politicians. Another approach aims to find bias words by employing IDF [211].

Another field that is relevant for the task of determining how persons are portrayed is sentiment analysis and more specifically target-dependent sentiment classification [212]. However, researchers have questioned whether the one-dimensional polarity scale of sentiment classification suffices to capture the actual fine-grained effects of framing (see Sect. 2.3.4). We will investigate sentiment classification and this question in Sect. 5.3.

5.2.2   Method

Given a person mention and its context, the objective of our method is to determine which frame properties the context as well as the mention yield on the person. Specifically, our simple method looks for words that express one or more frame properties on the person mention. Afterward, we aggregate for each target person frame properties from all its mentions in the given news article from sentence level to article level.[1]

The idea of our approach is to extend the one-dimensional polarity scale, i.e., positive, neutral, and negative, established in traditional sentiment classification with further classes, i.e., fine-grained properties, such as competent, powerful, and antonyms thereof. We call these fine-grained properties *frame properties*. Conceptually, some frame properties can be subsumed using sentiment polarity, but they also extend the characteristics that can be represented using polarity.

---

[1] The work described in this section was jointly conducted with our research for the method for context-driven cross-document coreference resolution. To avoid redundancy while improving readability, we refer to the respective sections in Chap. 4 and briefly summarize the work in the section at hand.

Table 5.1 Frame properties in NewsWCL50. Parentheses in the first column show the name of the respective antonym, if any

Name	# mentions	# mentions of antonym
Affection (refusal)	173	70
Trustworthiness (no trustworthiness)	43	120
Reason (unreason/irrationality)	72	84
Fairness / morality	6	
Confidence	65	
Easiness (difficulty)	2	99
Positive economy (negative economy)	24	35
Honor (dishonor)	30	17
Importance (unimportance)	242	15
Lawfulness (unlawfulness)	26	63
Power / leadership (weakness / passiveness)	517	173
Good quality / functioning (poor quality)	35	56
Aggressor (victim)	262	150
Safety (unsafety)	46	78
Positive (negative)	26	26
Other bias	172	

For example, while affection and refusal can be represented as more specific forms of positive and negative portrayal, respectively, other frame properties cannot be projected into the positive-negative scale. Such frame properties include, for example, (being portrayed as an) aggressor versus a victim, where the victim has neither positive nor negative sentiment polarity (the aggressor, though, has clearly negative polarity). This way, our approach seeks to overcome the shortcomings of the one-dimensional polarity scale used in sentiment classification (see Sect. 2.3.4).

Using a series of inductive and deductive small-scale content analyses, we devised in total 29 frame properties, of which 13 are bi-polar pairs and 3 have no antonym. Specifically, to derive the final set of frame properties, we initially asked coders to annotate any phrase that they felt was influencing their assessment of a person or other semantic concepts mentioned and also to state which perception, judgment, or feeling the phrase caused in them. We then used these initial open statements to derive a set of frame properties, which represented how the annotators felt a target was portrayed. Table 5.1 shows the final set of frame properties.

We refined these initial frame properties in an interactive process following best practices from the social sciences until three goals or factors were achieved or maximized. First, an acceptable inter-coder reliability was reached. Second, the set of frame properties covers a broad spectrum of person characteristics highlighted by "any" news coverage while, third, still being as specific as possible. The second and third goals aim to achieve a balance between being topic-independent and generic while preferring specific categories, such as "competent," over general categories, such as "positive." We achieved these goals after conducting six test iterations consisting of reviewing previous annotation results, refining the codebook and frame properties, and discussion of the changes with the annotators. We reached an acceptable inter-coder reliability of 0.65 (calculated with mean pairwise average agreement). The comparably low inter-coder reliability indicates the complexity and difficulty of the task (cf. [45]). Further information on the training prior to the main annotation is described in Sect. 4.3.2.2.

Finally, we conducted a deductive content analysis on 50 news articles reporting on policy issues using the codebook created during the training annotations. In the main annotation, 5926 mentions of persons and other target concepts were annotated. Further, 2730 phrases that each induce at least one frame property were annotated. For each frame property, additionally, the corresponding target concept had to be assigned. For example, in "Russia seizes Ukrainian naval ships," "Russia" would be annotated as a target concept of type "country," and "seizes" as a frame property with type "Aggressor" that targets "Russia." Each mention of a target concept in a text segment can be targeted by multiple frame property phrases. Further information on the training prior to the main annotation is described in Sect. 4.3.2.3.

After the annotation and in a one-time process, we manually defined a set of seed words for each of the frame properties $S_k \in S$. For each frame property S_k, we gathered seed words by carefully selecting common synonyms from a dictionary [233], e.g., for the frame property "affection," we selected the seed words: attachment, devotion, fondness, love, and passion.

For each news article passed to the frame analysis component, our method performs the following procedure. First, to identify frame property words, the method iterates all words in the given news article and determines for each word its semantic similarity to each of the frame properties. Specifically, we calculate the cosine similarity of the current word w and each seed word $s \in S_k$ of the current frame property S_k in a word embedding space [330]. We define the semantic similarity

$$\text{sim}(w, S_k) = \text{cossim}(\vec{w}, \vec{s}). \tag{5.1}$$

We assign to a word w any frame property S_k, where $\text{sim}(w, S_k) > t_p = 0.4$. At the end of this procedure, each word has a set of weighted frame properties. The weight of a frame property on a word is defined by $\text{sim}(w, S_k)$.

Second, for each target person c_i, we aggregate frame properties $S_k \in S$ from all its modifiers w_j of c_i found by dependency parsing [6]. We use manually devised rules to handle the different types of relations between head c_i and modifier w_j, e.g., to assign the frame properties of an attribute (modifier) to its noun (target person mention) or a predeterminer (modifier) to its head (target person mention).

Given a news article and a set of persons or other semantic concepts with one or more mentions, the output of the proposed method is as follows. For each mention, the method determines a set of weighted frame properties, yielded by the sentence of the mention. Further, for each semantic concept, the method returns a set of weighted frame properties by aggregating them from mention level to article level.

5.2.3 Exploratory Evaluation

We discuss the usability of this exploratory approach as to determining frame properties in a set of news articles reporting on the same event in two use cases. Due to the low inter-coder reliability of the frame property annotations (see Sect. 5.2.2), we expected low classification performance of our approach. Thus, we did not conduct a quantitative evaluation but instead qualitatively investigated the approach to derive future research ideas [46]. In contrast to the research objective of this thesis, i.e., identifying and communicating biases targeting persons, in this investigation, we also considered semantic concepts of the type groups of persons. This allows us to better demonstrate and discuss the results of the method.

In the first use case, we investigated the frame properties of persons and other semantic concepts in an event, where the DNC, a part of the Democratic Party in the USA, sued Russia and associates of Trump's presidential campaign in 2018 (see event #3 in Table 4.9). Table 5.2 shows exemplary frame properties of the three main actors involved in the event, Donald Trump, the Democratic Party, and the Russian Federation, each being a different concept type (shown in parentheses in Table 5.2). The first column shows each candidate's representative phrase (see Sect. 4.3.3.1). The linearly normalized scores $s(c, a, f)$ in the three exemplary frame property columns represent how strongly each article a (row) portrays a frame property f regarding a candidate c: $s = 1$ or -1 indicates the maximum presence of the property or its antonym, respectively. A value of 0 indicates the absence of the property or equal presence of the property and its antonym.

Left-wing outlets (LL and L) more strongly ascribe the property "aggressor" to Trump, e.g., $s(\text{Trump, LL, aggressor}) = 1$, than right-wing outlets do, for example, $s(\text{Trump, R, aggressor}) = 0.34$. This is conformal with the findings of manual analyses of news coverage of left- versus right-wing outlets regarding Republicans [71, 116, 117]. The Democratic Party is portrayed in all outlets as rather aggressive ($s = [0.91, 1]$), which can be expected due to the nature of the event, since the DNC sued various political actors.

The difficulty of frame property classification is visible in other frame properties that yielded inconclusive trends, such as "reason." We found that an increased level of abstractness is the main cause for lower frame identification performance (cf. [45, 211, 276]). For example, in the content analysis (see Sect. 4.3.2), we noticed that "reason" was often not induced by single words but rather more abstractly through actions that were assessed as reasonable by the human coders.

In the second use case, we investigated frame properties in an event where special counsel Mueller provided a list of questions to Trump in 2018 (see event #8 in Table 4.9). Table 5.3 shows selected frame properties of the two main actors involved in the event: Trump and Mueller. Since both are individual persons, their semantic concept type is "Actor." The results of our method indicated that the reviewed left-wing outlets ascribe rather positive frame properties to Mueller, e.g., $s(\text{Mueller, LL, confidence}) = 1$, than right-wing outlets do, $s(\ldots, \text{RR}, \ldots) = 0$. For Trump, we identified the opposite, e.g., $s(\text{Trump, LL, trustworthiness}) = -0.19$

Table 5.2 Excerpt of exemplary frame properties as determined automatically in the first use case

Concept label	Type	Outlet	honor	aggressor	reason
Trump	Actor	LL	-0.51	1.00	-0.32
		L	-0.75	0.76	0.00
		M	0.04	0.00	0.89
		R	0.00	0.34	0.00
		RR	0.00	0.44	0.00
Democratic Party	Group	LL	0.40	1.00	0.00
		L	0.57	1.00	0.00
		M	-1.00	0.91	-0.87
		R	0.93	1.00	-0.37
		RR	-0.98	1.00	0.00
Russia	Country	LL	1.00	0.53	0.00
		L	0.00	0.00	0.00
		M	0.00	-0.89	0.87
		R	1.00	0.03	0.00
		RR	-0.98	1.00	0.00

Table 5.3 Excerpt of exemplary frame properties as determined automatically in the second use case

Concept label	Type	Outlet	confidence	power	trustw.
Mueller	Actor	LL	0.70	0.00	1.00
		L	1.00	0.00	0.97
		M	0.61	0.00	0.63
		R	0.13	1.00	0.49
		RR	0.00	0.17	0.00
Trump	Actor	LL	-0.19	0.41	-0.93
		L	-0.80	0.00	-1.00
		M	0.41	0.00	-0.19
		R	0.48	0.00	0.00
		RR	1.00	0.47	-0.16

and $s(\ldots, RR, \ldots) = 1$. More strongly, left-wing news outlets even ascribe non-trustworthiness to Trump, e.g., $s(\text{Trump}, LL, \text{trustworthiness}) = -0.93$. Besides these expected patterns, other frame properties again showed inconclusive trends, such as power.

Due to the difficulty of automatically estimating frames (see Sect. 2.5), the identification of frame properties ascribed to persons and other semantic concepts did not always yield clear or expected patterns. We found this is especially true for abstract or implicitly ascribed frame properties. For example, we could not find clear patterns for the frame properties "reason" in the first use case (Table 5.2) and "power" in the second use case (Table 5.3), which is mainly due to the abstractness used to portray a person as being powerful or reasonable.

5.2.4 *Future Work*

In our exploratory evaluation, we found that our basic approach yielded trends concerning frame properties that are in part as expected and in part inconclusive. We think there are two main causes for the inconclusive results. First, the simplicity of the approach is one key reason. The second potential cause is the general difficulty of annotating or determining frames and, respectively, in our case the frames' effects [45].

To address the first issue, we propose to improve the approach using more sophisticated techniques, such as deep learning and recent language models. Fundamentally, our current approach is word-based and may often fail to catch the "meaning between the lines" (see Sect. 2.2.5). This is in contradiction to the substantial character of frames [80], which typically requires a higher degree of interpretation, being one key reason for the comparably "superficial" results of many automated approaches to date compared to manual content analyses (see Sect. 2.5). For example, determining implicitly ascribed frame properties, such as "reason," requires a high degree of interpretation since typically a news articles would not state that a person acted reasonably but instead this conclusion would be made by news consumers after reading one or more sentences describing, for example, actions that portray the person in a specific way. One idea to improve the classification performance is to use deep language models, such as RoBERTa, pre-trained on large amounts of also news articles [214]. RoBERTa and other language models have significantly improved natural language understanding capabilities across many tasks [373]. Given these advancements, we expect that such language models can also reliably determine complex and implicitly ascribed frame properties. However, fine-tuning language models, especially for multi-label classification tasks with high degree of required interpretation as for frames and frame properties, require also the creation of very large datasets (cf. [46]).

The second cause is the difficulty of frame annotation as well as frame classification or in our cases more specifically the effects of framing, i.e., frame properties. As our comparably low inter-coder reliability (see Sect. 5.2.2) and prior work indicate [45], the annotation of frames and frame properties is highly complex, and some "degree of subjectivity in framing analysis [is] unavoidable" [45]. In our view, the most effective idea to address the high annotation difficulty is to reduce the number of frame properties that are to be annotated. Other, commonly used means to tackle the subjectivity are performing more training iterations, might be less effective since we already conducted as many iterations as we could to improve the inter-coder reliability. Another promising idea is to determine frame properties on a much larger set of news articles. While our exploratory evaluation showed in principle also expected framing patterns, we tested our method only on five articles for each use case. We think that the task's ambiguity could—besides technical improvements as mentioned previously—be addressed by identifying framing patterns on more articles instead of attempting to pinpoint frame properties on individual articles.

5.2.5   *Conclusion*

This section presented the results of our exploratory research on imitating manual frame analysis. Albeit effective, such analysis entails the definition of topic-specific and analysis question-specific frames. Such dependencies are in contrast to the objectives of the person-oriented framing analysis approach, which is intended to be applied to any news coverage on policy issues. Instead of frames, we proposed to analyze frame properties, which represent the effects of person-oriented framing, such as whether a person is shown as being "aggressive."

In our view, the approach represents a promising line of research but at the same time suffers from shortcomings that are common to prior approaches aiming to imitate frame analysis, especially high annotation cost for the required training dataset. Likewise, we noticed a degree of subjectivity that could not be reduced without lowering the "substance" of the frame properties (cf. [45]), which is required to interpret the "meaning between the lines" like it is done in frame analysis.

The dataset used for the approach and its codebook are available at https://github.com/fhamborg/NewsWCL50.

5.3   **Target-Dependent Sentiment Classification**

This section describes the second approach for our frame analysis component. Specifically, we describe a dataset and method for target-dependent sentiment classification (TSC, also called aspect-term sentiment classification) in news articles. In the context of the overall person-oriented framing analysis (PFA) and in particular the frame analysis component, we use TSC to classify a fundamental effect of person-oriented framing, i.e., whether sentences and articles portray individual persons positively or negatively. As we show in this section and our prototype evaluation (Chap. 6), TSC represents a pragmatic and effective alternative to the fine-grained but expensive approach of classifying frame properties. The advantages of TSC over approaches aiming to capture frames or frame derivatives are the reduced annotation cost and high reliability.

We define our objective in this section as follows: we seek to detect polar judgments toward target persons [335]. Following the TSC literature, we include only in-text, specifically in-sentence, means to express sentiment. In news texts, such means are, for example, word choice or describing actions performed by the target, e.g., "John and Bert got in a fight" or "John attacked Bert." Sentiment can also be expressed indirectly, e.g., through quoting another person, such as "According to John, an expert on the field, the idea 'suffers from fundamental issues' such as [. . .]" [335]. Other means may also alter the perception of persons and topics in the news, but are not in the scope of the task [16], e.g., because they are not on sentence level, for example, story selection, source selection, article's placement and size (Sect. 2.1), and epistemological bias [297]. Albeit excluding non-sentence-level means from our objective in this section, in the context of the overall thesis, the TSC method will still be able to catch the effects of source selection and commission and omission of information. For example, when journalists write articles and include mostly information of one perspective that is in favor of specific persons, the resulting article will mostly reflect that perspective and thus be in favor of these persons (Sect. 3.3.2).

The main contributions of this section are as follows: (1) We create a small-scale dataset and train state-of-the-art models on it to explore characteristics of sentiment in news articles. (2) We introduce *NewsMTSC*, a large, manually annotated dataset for TSC in political news articles. We analyze the quality and characteristics of the dataset using an on-site, expert annotation. Because of its fundamentally different characteristics compared to previous TSC datasets, e.g., as to how sentiment is expressed and text lengths, NewsMTSC represents a challenging novel dataset for the TSC task. (3) We propose a neural model that improves TSC performance on news articles compared to prior state-of-the-art models. Additionally, our model yields competitive performance on established TSC datasets. (4) We perform an extensive evaluation and ablation study of the proposed model. Among others, we investigate the recently claimed "degeneration" [161] of TSC to sequence-level classification, finding a performance drop in all models when comparing single- and multi-target sentences.

The remainder of this section is structured as follows. In Sect. 5.3.1, we provide an overview of related work and identify the research gap of sentiment classification in news articles. In Sect. 5.3.2, we explore the characteristics of how sentiment is expressed in news articles by creating and analyzing a small-scale TSC dataset. We then use and address the findings of this exploratory work, to create our main dataset (Sect. 5.3.3) and model (Sect. 5.3.4). Key differences and improvements of the main dataset compared to the small-scale dataset are as follows. We significantly increase the dataset's size and the number of annotators per example and address class imbalance. Further, we devise annotation instructions specifically created to capture a broad spectrum of sentiment expressions specific to news articles. In contrast, the early dataset misses the more implicit sentiment expressions commonly used by news authors (see Sect. 5.3.2.5). Also, we test various consolidation strategies and conduct an expert annotation to validate the dataset.

We provide the dataset and code to reproduce our experiments at https://github.com/fhamborg/NewsMTSC.

5.3.1 Related Work

Analogously to other NLP tasks, the TSC task has recently seen a significant performance leap due to the rise of language models [73]. Pre-BERT approaches yield up to $F1_m = 63.3$ on the SemEval 2014 Twitter set [182]. They employ traditional machine learning combining hand-crafted sentiment dictionaries, such as

SentiWordNet [13], and other linguistic features [29]. On the same dataset, vanilla BERT (also called BERT-SPC) yields 73.6 [73, 400]. Specialized downstream architectures improve performance further, e.g., LCF-BERT yields 75.8 [400].

The vast majority of recently proposed TSC approaches employ BERT and focus on devising specialized downstream architectures [329, 346, 400]. More recently, to improve performance further, additional measures have been proposed, for example, domain adaption of BERT, i.e., domain-specific language model fine-tuning prior to the TSC fine-tuning [76, 300]; use of external knowledge, such as sentiment or emotion dictionaries [151, 401], rule-based sentiment systems [151], and knowledge graphs [102]; use of all mentions of a target and/or related targets in a document [50]; and explicit encoding of syntactic information [286, 398].

To train and evaluate recent TSC approaches, three datasets are commonly used: Twitter [257, 258, 305], Laptop, and Restaurant [289, 290]. These and other TSC datasets [273] suffer from at least one of the following shortcomings. First, implicitly or indirectly expressed sentiment is rare in them. In their domains, e.g., social media and reviews, typically, authors explicitly express their sentiment regarding a target [402]. Second, they largely neglect that a text may contain coreferential mentions of the target or mentions of different concepts (with potentially different polarities), respectively [161].

Texts in news articles differ from reviews and social media in that news authors typically do not express sentiment toward a target explicitly (exceptions include opinion pieces and columns). Instead, journalists implicitly or indirectly express sentiment (Sect. 2.3.4) because language in news is typically expected to be neutral and journalists to be objective [16, 110].

Our objective as described in the beginning of Sect. 5.3 is largely identical to prior news TSC literature [16, 335] with key differences: we do not generally discard the "author level" and "reader level." Doing so would neglect large parts of sentiment expressions. Thus, it would degrade real-world performance of the resulting dataset and models trained on it. For example, word choice (listed as "author level" and discarded from their problem statement) is in our view an in-text means that may in fact strongly influence how readers perceive a target, e.g., "compromise" or "consensus." While we do not exclude the "reader level," we do seek to exclude polarizing or contentious cases, where no uniform answer can be found in a set of randomly selected readers (Sects. 5.3.3.3 and 5.3.3.4). As a consequence, we generally do not distinguish between the three levels of sentiment ("author," "reader," and "text").

Previous news TSC approaches mostly employ sentiment dictionaries, e.g., created manually [16, 335] or extended semi-automatically [110], but yield poor or even "useless" [335] performances. To our knowledge, there exist two datasets for the evaluation of news TSC methods. Steinberger et al. [335] proposed a news TSC dataset, which—perhaps due to its small size ($N = 1274$)—has not been used or tested in recent TSC literature. Another dataset contains quotes extracted from news articles, since quotes more likely contain explicit sentiment ($N = 1592$) [16].

In summary, no suitable datasets for news TSC exist nor have news TSC approaches been proposed that exploit recent advances in NLP.

5.3.2 *Exploring Sentiment in News Articles*

We describe how the procedure used to create our exploratory TSC dataset for the domain of news articles, including the collection of articles and the annotation procedure. Afterward, we discuss the characteristics of the dataset. Then, we report the results of our evaluation where we test state-of-the-art TSC models on the dataset. Lastly, we discuss the findings and shortcomings of our qualitative dataset investigation and quantitative evaluation to derive means to address these shortcomings in our main dataset.

5.3.2.1 Creating an Exploratory Dataset

Our procedure to create the exploratory dataset for sentiment classification on news articles entails the following steps. First, we create a base set of articles of high diversity in topics covered and writing styles, e.g., whether emotional or factual words are used (cf. [96]). Using our news extractor (Sect. 3.5), we collect news articles from the Common Crawl news crawl (CCNC, also known as CC-NEWS), consisting of over 250M articles until August 2019 [256]. To ensure diversity in writing styles, we select 14 US news outlets,[2] which are mostly major outlets that represent the political spectrum from left to right, based on selections by Budak et al. [38], Groseclose and Milyo [117], and Baum and Groeling[21]. We cannot simply select the whole corpus, because CCNC lacks articles for some outlets and time frames. By selecting articles published between August 2017 and July 2019, we minimize such gaps while covering a time frame of 2 years, which is sufficiently large to include many diverse news topics. To facilitate the balanced contribution of each outlet and time range, we perform binning: we create 336 bins, one for each outlet and month, and randomly draw 10 articles reporting on politics for each bin, resulting in 3360 articles in total.[3] During binning, we remove any article duplicates by text equivalence.

To create examples for annotation, we select all mentions of NEs recognized as PERSON, NROP, or ORG for each article [376].[4] We discard NE mentions in sentences shorter than 50 characters. For each NE mention, we create an example by using the mention as the target and its surrounding sentence as its context. We remove any example duplicates. Afterward, to ensure diversity in writing styles and

---

[2] BBC, Breitbart, Chicago Tribune, CNN, Los Angeles Daily News, Fox News, HuffPost, Los Angeles Times, NBC, The New York Times, Reuters, USA Today, The Washington Post, and The Wall Street Journal.

[3] To classify whether an article reports on politics, we use a DistilBERT-based [310] classifier with a single dense layer and softmax trained on the HuffPost [244] and BBC datasets [115]. During the subsequent manual annotation, coders discard remaining, non-political articles.

[4] For this task, we use spaCy v2.1: https://spacy.io/usage/v2-1.

topics, we use the outlet-month binning described previously and randomly draw examples from each bin.

Different means may be used to address expected class imbalance, e.g., for the Twitter set, only examples that contained at least one word from a sentiment dictionary were annotated [257, 258]. While doing so yields high frequencies of classes that are infrequent in real-world distribution, it also causes dataset shift and selection bias [293]. Thus, we instead investigate the effectiveness of different means to address class imbalance during training and evaluation (see Sect. 5.3.2.4).

5.3.2.2 Annotating the Exploratory Dataset

We set up an annotation process following best practices from TSC literature [258, 290, 305, 335]. For each example, we asked three coders to read the context, in which we visually highlighted the target, and assess the target's sentiment. Examples were shown in random order to each coder. Coders could choose from *positive*, *neutral*, and *negative* polarity, whereby they were allowed to choose positive and negative polarity at the same time. Coders were asked to *reject* an example, e.g., if it was not political or a meaningless text fragment. Before, coders read a codebook that included instructions on how to code and examples. Five coders, students, aged between 24 and 32, participated in the process.

In total, 3288 examples were annotated, from which we discard 125 (3.8%) that were rejected by at least one coder, resulting in 3163 non-rejected examples. From these, we discard 3.3% that lacked a majority class, i.e., examples where each coder assigned a different sentiment class, and 1.8% that were annotated as positive and negative sentiment at the same time, to allow for better comparison with previous TSC datasets and methods (see Sect. 5.3.1). Lastly, we split the remaining 3002 examples into training and test sets; see Table 5.4.

We use the full set of 3163 non-rejected examples to illustrate the degree of agreement between coders: 3.3% lack a majority class; for 62.7%, two coders assigned the same sentiment; and for 33.9%, all coders agreed. On average, the accuracy of individual coders is $A_h = 72.9\%$. We calculate two inter-rater reliability (IRR) measures. For completeness, Cohen's kappa is $\kappa = 25.1$, but it is unreliable in our case due to Kappa's sensitivity to class imbalance [55]. The mean pairwise observed agreement over all coders is 72.5.

Table 5.4 Class frequencies of the splits in the exploratory TSC dataset

Set	Negative	Neutral	Positive	Total
training	530	1600	171	2301
test	167	487	47	701
total	697	2087	218	3002

5.3.2.3 Exploring the Characteristics of Sentiment in News Articles

In a manual, qualitative analysis of our exploratory dataset, we found two key differences of news compared to established domains: First, we confirmed that news contains mostly implicit and indirect sentiment (see Sect. 5.3.1). Second, determining the sentiment in news articles typically requires a greater degree of interpretation (cf. [335]). The second difference is caused by multiple factors, particularly the implicitness of sentiment (mentioned as the first difference) and that sentiment in news articles is more often dependent on non-local, i.e., off-sentence, context. In the following, we discuss annotated examples (part of the dataset and discarded examples) to understand the characteristics of target-dependent sentiment in news texts.

In our analysis, we found that in news articles, a key means to express targeted sentiment is to describe actions performed by the target. This is in contrast, e.g., to product reviews where more often a target's feature, e.g., "high resolution," or the mention of the target itself, e.g., "the camera is awesome," expresses sentiment. For example, in "The Trump administration has worked tirelessly to impede a transition to a green economy with actions ranging from opening the long-protected Arctic National Wildlife Refuge to drilling [. . .]," the target (underlined) was assigned negative sentiment due to its actions.

We found sentiment in ≈3% of the examples to be strongly reader-dependent (cf. [16]).[5] In the previous example, the perceived sentiment may, in part, depend on the reader's own ideological or political stance, e.g., readers focusing on economic growth could perceive the described action positively, whereas those concerned with environmental issues would perceive it negatively.

In some examples, targeted sentiment expressions can be interpreted differently due to ambiguity. As a consequence, we mostly found such examples in the discarded examples, and thus they are not contained in our exploratory dataset. While this can be true for any domain (cf. "polarity ambiguity" in [290]), we think it is especially characteristic for news articles, which are lengthier than tweets and reviews, giving authors more ways to refer to non-local statements and to embed their arguments in larger argumentative structures. For instance, in "And it is true that even when using similar tactics, President Trump and President Obama have expressed very different attitudes towards immigration and espoused different goals," the target was assigned neutral sentiment. However, when considering this sentence in the context of its article [356], the target's sentiment may be shifted (slightly) negatively.

From a practical perspective, considering more context than only the current sentence seems to be an effective means to determine otherwise ambiguous sentiment expressions. By considering a broader context, e.g., the current sentence and previous sentences, annotators can get a more comprehensive understanding of the

---

[5] We drew a random sample of 300 examples and concluded in a two-person discussion that the sentiment in 8 examples could be perceived differently.

author's intention and the sentiment the author may have wanted to communicate. The greater degree of interpretation required to determine non-explicit sentiment expressions may naturally lead to a higher degree of subjectivity. Due to our majority-based consolidation method (see Sect. 5.3.2.2), examples with non-explicit or apparently ambiguous sentiment expressions are not contained in our exploratory dataset.

5.3.2.4   Experiments and Discussion

We evaluated three TSC methods that represent the state of the art on the established TSC datasets Laptop, Restaurant, and Twitter: AEN-BERT [329], BERT-SPC [73], and LCF-BERT [400]. Additionally, we tested the methods using a domain-adapted language model, which we created by fine-tuning BERT (base, uncased) for 3 epochs on 10M English sentences sampled from CCNC (cf. [300]). For all methods, we test hyperparameter ranges suggested by their respective authors.[6] Additionally, we investigated the effects of two common measures to address class imbalance: weighted cross-entropy loss (using inverse class frequencies as weights) and oversampling of the training set. Of the training set, we use 2001 examples for training and 300 for validation.

We used average recall (R_a) as our primary measure, which was also chosen as the primary measure in the TSC task of the latest SemEval series, due to its robustness against class imbalance [305]. We also measured accuracy (A), macro F1 ($F1_m$), and average F1 on positive and negative classes ($F1_{pn}$) to allow comparison to previous works [258].

Table 5.5 shows that LCF-BERT performed best ($R_a = 67.3$ using BERT and 69.8 using our news-adapted language model).[7] Class-weighted cross-entropy loss helped best to address class imbalance ($R_a = 69.8$ compared to 67.2 using oversampling and 64.6 without any measure).

Performance in news articles was significantly lower than in established domains, where the top model (LCF-BERT) yielded in our experiments $R_a = 78.0$ (Laptop), 82.2 (Restaurant), and 75.6 (Twitter). For Laptop and Restaurant, we used domain-adapted language models [300]. News TSC accuracy $A = 66.0$ was lower than single human level $A_h = 72.9$ (see Sect. 5.3.2.3).

We carried out a manual error analysis (up to 30 randomly sampled examples for each true class). We found *target misassociation* as the most common error cause: In 40%, sentences express the predicted sentiment toward a different target. In 30%,

---

[6] Epochs $\in \{3, 4\}$; batch size $\in \{16, 32\}$; learning rate $\in \{2e - 5, 3e - 5, 5e - 5\}$; label smoothing regularization (LSR) [353], $\epsilon \in \{0, 0.2\}$; dropout rate, 0.1; $\mathcal{L}_2$ regularization, $\lambda = 10^{-5}$. We used Adam optimization [181], Xavier uniform initialization [109], and cross-entropy loss [113]. Where multiple values for a hyperparameter are given, we tested all their combinations in an exhaustive search.

[7] Each row in Table 5.5 shows the results of the hyperparameters that performed best on the validation set.

Table 5.5 TSC performance on the exploratory dataset. *LM* refers to the language model used, where *base* is BERT (base, uncased) and *news* is our fine-tuned BERT model

LM	Method	R_a	A	$F1_m$	$F1_{pn}$
base	AEN-BERT	59.7	62.9	55.0	47.3
	BERT-SPC	62.1	62.1	53.3	44.9
	LCF-BERT	67.3	61.3	54.4	46.5
news	AEN-BERT	59.8	62.9	54.5	46.2
	BERT-SPC	66.7	63.5	55.0	45.8
	LCF-BERT	**69.8**	66.0	58.8	51.4

we cannot find any apparent cause. The remaining cases contain various potential causes, including usage of euphemisms or sayings (12% of examples with negative sentiment). Infrequently, we found that sentiment is expressed by rare words or figurative speech or is reader-dependent (the latter in 2%, approximately matching the 3% of reader-dependent examples reported in Sect. 5.3.2.3).

Previous news TSC approaches, mostly dictionary-based, could not reliably classify implicit or indirect sentiment expressions (see Sect. 5.3.1). In contrast, our experiments indicate that BERT's language understanding suffices to interpret implicitly expressed sentiment correctly (cf. [16, 73, 110]). Our exploratory dataset does not contain instances in which the broader context defines sentiment, since human coders could or did not classify them in our annotation procedure. Our experiments therefore cannot elucidate this particular characteristic discussed in Sect. 5.3.2.3.

5.3.2.5 Summary

We explored how target-dependent sentiment classification (TSC) can be applied to political news articles. After creating an exploratory dataset of 3000 manually annotated sentences sampled from news articles reporting on policy issues, we qualitatively analyzed its characteristics. We found notable differences concerning how authors express sentiment toward targets as compared to other, well-researched domains of TSC, such as product reviews or posts on social media. In these domains, authors tend to explicitly express their opinions. In contrast, in news articles, we found dominant use of implicit or indirect sentiment expressions, e.g., by describing actions, which were performed by a given target, and their consequences. Thus, sentiment expressions may be more ambiguous, and determining their polarity requires a greater degree of interpretation.

In our quantitative evaluation, we found that current TSC methods performed lower on the news domain (average recall $R_a = 69.8$ using our news-adapted BERT model, $R_a = 67.3$ without) than on popular TSC domains ($R_a = [75.6, 82.2]$).

While our exploratory dataset contains clear sentiment expressions, it lacks other sentiment types that occur in real-world news coverage, for example, sentences that express sentiment more implicitly or ambiguously. To create a labeled TSC dataset that better reflects real-world news coverage, we suggest to adjust annotation instructions to raise annotators' awareness of these sentiment types and clearly define how they should be labeled. Technically, apparently ambiguous sentiment expressions might be easier to label when considering a broader context, e.g., not only the current sentence but also previous sentences. Considering more context might also help to improve a classifier's performance.

5.3.3   NewsMTSC: Dataset Creation

This section describes the procedure to create our main dataset for TSC in the news domain. When creating the dataset, we rely on best practices reported in literature on the creation of datasets for NLP [291], especially for the TSC task [305]. As our previous exploration has showed (Sect. 5.3.2.5), compared to previous TSC datasets though, the nature of sentiment in news articles requires key changes, especially in the annotation instructions and consolidation of answers [335].

5.3.3.1   Data Sources

We use two datasets as sources: our POLUSA dataset [96] and the Bias Flipper 2018 (BF18) dataset [49]. Both satisfy five criteria that are important to our problem. First, they contain news articles reporting on political topics. Second, they approximately match the online media landscape as perceived by an average US news consumer.[8] Third, they have a high diversity in topics due to the number of articles contained and time frames covered (POLUSA: 0.9M articles published between Jan. 2017 and Aug. 2019, BF18: 6447 articles associated with 2781 events). Fourth, they feature high diversity in writing styles because they contain articles from across the political spectrum, including left- and right-wing outlets. Fifth, we find that they contain only few minor content errors albeit being created through scraping or crawling.

In early tests when selecting data sources, we tested other datasets as well. While we found that other factors are more important for the resulting quality of annotated examples (filtering of candidate example, annotation instructions, and consolidation strategy), we also found that other datasets are slightly less suitable as to the five previously mentioned criteria because the datasets, e.g., contain only contentious news topics and articles [45] or hyperpartisan sentences [178], are of mixed content quality [264] or contain too few sentences [4, 5].

---

[8] Each dataset roughly approximates the US media landscape, i.e., POLUSA by design [96], and BF18 because it was crawled from a news aggregator on a daily basis [49].

5.3.3.2 Creation of Examples

To create a batch of examples for annotation, we devise a three tasks process: First, we extract example candidates from randomly selected articles. Second, we discard non-optimal candidates. Only for the train set, third, we filter candidates to address class imbalance. We repeatedly execute these tasks so that each batch yields 500 examples for annotation, contributed equally by both sources.

First, we randomly select articles from the two sources. Since both are at least very approximately uniformly distributed over time [49, 96], randomly drawing articles will yield sufficiently high diversity in both writings styles and reported topics (Sect. 5.3.3.1). To extract from an article examples that contain meaningful target mentions, we employ coreference resolution (CR).[9] We iterate all resulting coreference clusters of the given article and create a single example for each mention and its enclosing sentence.

Extraction of mentions of named entities (NEs) is the commonly employed method to create examples in previous TSC datasets [257, 258, 305, 335]. We do not use it since we find it would miss $\gtrsim$30% mentions of relevant target candidates, e.g., pronominal or near-identity mentions.

Second, we perform a two-level filtering to improve quality and "substance" of candidates. On coreference cluster level, we discard a cluster c in a document d if $|M_c| \leq 0.2|S_d|$, where $|\ldots|$ is the number of mentions of a cluster (M_c) and sentences in a document (S_d). Also, we discard non-persons clusters, i.e., if $\exists m \in M_c : t(m) \notin \{-, P\}$, where $t(m)$ yields the NE type[10] of m and $-$ and P represent the unknown and person type, respectively. On example level, we discard short and similar examples e, i.e., if $|s_e| < 50 \lor \exists \hat{e} : \text{sim}(s_e, s_{\hat{e}}) > 0.6 \land m_e = m_{\hat{e}} \land t_e = t_{\hat{e}}$ where s_e, m_e, and t_e are the sentence of e, its mention, and the target's cluster, respectively, and $\text{sim}(\ldots)$ is the cosine similarity. Lastly, if a cluster has multiple mentions in a sentence, we try to select the most meaningful example. In short, we prefer the cluster's representative mention[11] over nominal mentions and those over all other instances.

Third, for only the train set, we filter candidates to address class imbalance. Specifically, we discard examples e that are likely the majority class ($p(\text{neutral}|s_e) > 0.95$) as determined by a simple binary classifier [310]. Whenever annotated and consolidated examples are added to the train set of NewsMTSC, we retrain the classifier on them and all previous examples in the train set.

---

[9] We employ spaCy 2.1 (https://github.com/explosion/spaCy/releases/tag/v2.1.8) and neuralcoref 4.0 (https://github.com/huggingface/neuralcoref/releases/tag/v4.0.0).

[10] Determined by spaCy.

[11] Determined by neuralcoref.

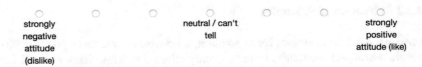

strongly   neutral / can't   strongly
negative   tell   positive
attitude   attitude (like)
(dislike)

Fig. 5.1 Final version of the annotation instructions as shown on Amazon Mechanical Turk

5.3.3.3   Annotation

Instructions used in popular TSC datasets plainly ask annotators to rate the sentiment of a text toward a target [290, 305]. For news texts, we find that doing so yields two issues [16]: low inter-rater reliability (IRR) and low suitability. Low suitability refers to examples where annotators' answers can be consolidated but the resulting majority answer is incorrect as to the task. For example, instructions from prior TSC datasets often yield low suitability for polarizing targets, independent of the sentence they are mentioned in. Figure 5.1 depicts our final annotation instructions.

In an interactive process with multiple test annotations (six on-site and eight on Amazon Mechanical Turk, MTurk), we test various measures to address the two issues. We find that asking annotators to think from the perspective of the sentence's author strongly facilitates that annotators overcome their personal attitude. Further, we find that we can effectively draw annotators' attention not only at the event and other "facts" described in the sentence (the "what") but also at word choice ("how" it is described) by exemplarily mentioning both factors and abstracting these factors as the author's holistic "attitude."[12] We further improve IRR and suitability, e.g., by explicitly instructing annotators to rate sentiment only regarding the target but not regarding other aspects, such as the reported event.

We also test other means to address low IRR and suitability in news TSC annotation but find our means to be more efficient while similarly effective. For example, Balahur et al. [16] ask annotators to only rate the target's sentiment but not consider whether the news are "good" or "bad." They also ask annotators to interpret only "what is said" and not use their own background knowledge. Additionally, we test a design where we ask annotators to select the more negative sentence of a pair of sentences sharing a target. We use semantic textual similarity (STS) datasets [4, 5] and extract all pairs with an STS score >2.5. While this design yields high IRR, suitability (especially political framing through word choice is found more effectively [166]), and efficiency, the STS datasets contain too few examples. On MTurk, we find consistently across all instruction variants that short instructions yield higher suitability and IRR than more comprehensive instructions. Surprisingly, the average duration of each crowdworkers' first assignment is shorter for the latter. This is perhaps because crowdworkers have high incentive to minimize the duration

---

[12] To think from the author's perspective is not to be confused with the "author level" defined by Balahur et al. [16].

per task to increase their salary and in case they deem instructions too long, the crowdworkers will not read them at all or only very briefly [302, 322].

While most TSC dataset creation procedures use 3- or 5-point Likert scales [16, 257, 258, 289, 290, 305, 335], we use a 7-point scale to encourage rating also only slightly positive or negative examples as such.

Technically, we closely follow previous literature on TSC datasets [290, 305]. We conduct the annotation of our examples on MTurk. Each example is shown to five randomly selected crowdworkers. To participate in our annotation, crowdworkers must have the "Master" qualification, i.e., have a record of successfully completed, high-quality work on MTurk. To ensure quality, we implement a set of objective measures and tests [180]. While we pay all crowdworkers always (USD 0.07 per assignment), we discard all of a crowdworker's answers if at least one of the following conditions is met. (a) A crowdworker was not shown any test question or answered at least one incorrectly,[13] (b) a crowdworker provided answers to invisible fields in the HTML form (0.3% of crowdworkers did so, supposedly bots), or (c) the average duration of time spent on the assignments was extremely low ($<4s$).

The IRR is sufficiently high ($\kappa_C = 0.74$) when considering only examples in NewsMTSC. The expected mixed quality of crowdsourced work becomes apparent when considering all examples, including those that could not be consolidated and answers of those crowdworkers who did not pass our quality checks ($\kappa_C = 0.50$).

5.3.3.4   Consolidation

We consolidate the answers of each example to a majority answer by employing a restrictive strategy. Specifically, we consolidate the set of five answers A to the single-label three-class polarity $p \in \{\text{pos., neu., neg.}\}$ if $\exists C \subseteq A : |C| \geq 4 \wedge \forall c \in C : s(c) = p$, where $s(c)$ yields the three-class polarity of an individual seven-class answer c, i.e., neutral $\Rightarrow$ neutral, any positive (from slightly to strongly) $\Rightarrow$ positive, and, respectively, for negative. If there is no such consolidation set C, A cannot be consolidated, and the example is discarded. Consolidating to three-class polarity allows for direct comparison to established TSC dataset.

While the strategy is restrictive (only 50.6% of all examples are consolidated this way), we find it yields the highest quality. We quantify the dataset's quality by comparing the dataset to an expert annotation (Sect. 5.3.3.6) and by training and testing models on dataset variants with different consolidations. Compared to consolidations employed for previous TSC datasets, quality is improved significantly on our examples, e.g., our strategy yields $F1_m = 86.4$ when compared to experts' annotations and models trained on the resulting set yield up to $F1_m = 83.1$, whereas the two-step majority strategy employed for the Twitter 2016 set [258] yields 50.6 and 53.4, respectively.

---

[13] Prior to submitting a batch of examples to MTurk, we add 6% test examples with unambiguous sentiment, e.g., "Mr. Smith is a bad guy."

Table 5.6 Class frequencies of NewsMTSC. Columns (f.l.t.r.): name; count of targets with any, positive, neutral, and negative sentiment, respectively; and count of examples with multiple targets of any and different polarity, respectively

Set	Total	Pos.	Neu.	Neg.	MT-a	MT-d
Train	8739	2395	3028	3316	972	341
Test-mt	1476	246	748	482	721	294
Test-rw	1572	361	587	624	73	30

5.3.3.5   Splits and Multi-Target Examples

NewsMTSC consists of three sets as depicted in Tables 5.6 and 5.7. For the *train* set, we employ class balancing prior to annotation (Sect. 5.3.3.2). To minimize dataset shift, which might yield a skewed sentiment distribution in the dataset compared to the real world [293], we do not use class balancing for either of the two test sets. Sentences can have multiple targets (MT) with potentially different polarities. We call this *MT property*. To investigate the effect on TSC performance of considering or neglecting the MT property [161], we devise a test set named *test-mt*, which consists only of examples that have at least two semantically different targets, i.e., each belonging to a separate coreference cluster (Sect. 5.3.3.2). Since the additional filtering required for *test-mt* naturally yields dataset shift, we create a second test set named *test-rw*, which omits the MT filtering and is thus designed to be as close as possible to the real-world distribution of sentiment. We seek to provide a sentiment score for each person in each sentence in *train* and *test-rw*, but mentions may be missing, e.g., because of erroneous coreference resolution or crowdworkers' answers could not be consolidated. Table 5.7 shows the frequencies of the targets and sentiment classes with added coreferential mentions.

Table 5.7 Statistics of coreference-related examples in NewsMTSC. Columns (f.l.t.r.): name and count of targets and their coreferential mentions with any, positive, neutral, and negative sentiment, respectively

Set	+Corefs	Pos.	Neu.	Neg.
Train	11880	3434	3744	4702
Test-mt	1883	333	910	640
Test-rw	1572	361	587	624

5.3.3.6 Quality and Characteristics

We conducted an expert annotation of a random subset of 360 examples used during the creation of NewsMTSC with 5 international graduate students (studying Political or Communication Science at the University of Zurich, Switzerland, 3 female, 2 male, aged between 23 and 29). Key differences compared to the MTurk annotation are as follows: First is extensive training until high IRR is reached (considering all examples, $\kappa_C = 0.72$; only consolidated, $\kappa_C = 0.93$). We conducted five iterations, each consisting of individual annotations by the students, quantitative and qualitative review, adaption of instructions, and individual and group discussions. Second are comprehensive instructions (4 pages). Third is no time pressure, since the students were paid per hour (crowdworkers per assignment).

When comparing the expert annotation with our dataset, we found that NewsMTSC is of high quality ($F1_m = 86.4$). The quality of unfiltered answers from MTurk is, as expected, much lower (50.1).

What is contained in NewsMTSC? In a random set of 50 consolidated examples from MTurk, we found that most frequent, non-mutually exclusive means to express a polar statement (62% of the 50) are usage of quotes (in total, direct, and indirect 42%, 28%, and 14%, respectively), target being subject to action (24%), evaluative expression by the author or an opinion holder mentioned outside of the sentence (18%), target performing an action (16%), and loaded language or connoted terms (14%). Direct quotes often contain evaluative expressions or connoted terms and indirect quotes less. Neutral examples (38% of the 50) contain mostly objective storytelling about neutral events (16%) or variants of "[target] said that [. . .]" (8%). Yet, "said" variants cannot be used as a reliably indicator for neutral sentiment, e.g., if the target has multiple mentions in the sentences or if the target's statement is considered positive or negative, e.g., "'Not all of that is preventable, but a lot of it is preventable if we've got better cooperation [. . .],' Obama said."

What is not contained in NewsMTSC? We qualitatively reviewed all examples where individual answers could not be consolidated to identify potential causes why annotators do not agree. The predominant reason is technical, i.e., the restrictiveness of the consolidation (MTurk compared to experts: 26% $\approx$ 30%). Other examples lack apparent causes (24% $\gg$ 8%). Further potential causes are (not mutually exclusive) as follows: ambiguous sentence (16% $\approx$ 18%), sentence contains positive and negative parts (8% $\approx$ 6%), and opinion holder is target (6% $\approx$ 8%), e.g., "[. . .] Bauman asked supporters to 'push back' against what he called a targeted campaign to spread false rumors about him online." In a subset of such instances, more context could have helped to resolve ambiguity, e.g., by showing annotators also the sentence prior to the mention.

What are qualitative differences in the annotations by crowdworkers and experts? We reviewed all 63 cases (18%) where answers from MTurk could be consolidated but differ to experts' answers. The major reason for disagreement is the restrictiveness of the consolidation (53 cases have no consolidation among the experts). In ten cases, the consolidated answers differ. We found that in few examples (2–3%), crowdsourced annotations are superficial and fail to interpret the full sentence correctly.

Texts in NewsMTSC are much longer than in prior TSC datasets (mean over all examples): 152 characters compared to 100, 96, and 90 in Twitter, Restaurant, and Laptops, respectively.

5.3.4 Method

The goal of TSC is to find a target's polarity $y \in \{$pos., neu., neg.$\}$ in a sentence. Our model consists of four key components (Fig. 5.2): a pre-trained language model (LM), a representation of external knowledge sources (EKS), a target mention mask, and a bidirectional GRU (BiGRU) [51]. We adapt our model from Hosseinia, Dragut, and Mukherjee [151] and change the design as follows: we employ a target mask (which they did not) and use multiple EKS simultaneously (instead of one). Further, we use a different set of EKS (Sect. 5.3.5) and do not exclude the LM's parameters from fine-tuning.

Input Representation

We construct three model inputs. The first is a text input T constructed as suggested by Devlin et al. [73] for question answering (QA) tasks. Specifically, we concatenate the sentence and target mention and tokenize the two segments using the LM's tokenizer and vocabulary, e.g., WordPiece for BERT [386].[14] This step results in a text input sequence $T = [\text{CLS}, s_0, s_1, \ldots, s_p, \text{SEP}, t_0, t_1, \ldots, t_q, \text{SEP}] \in \mathbb{N}^n$ consisting of n word pieces, where n is the manually defined maximum sequence length.

Various forms of this representation have been proposed, e.g., opposite order sentence and target or instead of the plain target mention using a natural language question or pseudo sentence [151, 346]. We find that on average in the TSC domain, they yield lower performance than the plain variant that we employ.

The second input is a feature representation of the sentence, which we create using one or more EKS, such as dictionaries [151, 401]. Given an EKS with d dimensions, we construct an EKS representation $E \in \mathbb{R}^{n \times d}$ of S, where each vector $e_{i \in \{0, 1, \ldots, p\}}$ is a feature representation of the word piece i in the sentence. For example, when using a sentiment dictionary with two mutually non-exclusive polarities' dimensions positive and negative [153], $d = 2$. Given a sentence "Good [. . .]," we set $e_1 = [1, 0]$. To facilitate learning associations between the token-based EKS representation and the WordPiece-based sequence T, we create E so that it contains k repeated vectors for each token where k is the token's number of word pieces. Thereby, we also consider special characters, such as CLS. If multiple

---

[14] For readability, we showcase inputs as used for BERT. We adapt inputs correspondingly for other LMs.

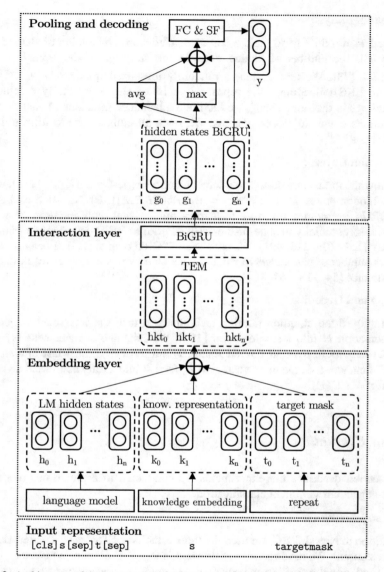

Fig. 5.2 Architecture of the proposed model for target-dependent sentiment classification

EKS with a total number of dimensions $\hat{d} = \sum d$ are used, their representations of the sentence are stacked resulting in $E \in \mathbb{R}^{n \times \hat{d}}$.

The third input is a target mask $M \in \mathbb{R}^n$, i.e., for each word piece i in the sentence that belongs to the target, $m_i = 1$, else 0 [94].

Embedding Layer

We feed T into the LM to yield a contextualized word embedding of shape $\mathbb{R}^{n \times h}$, where h is the number of hidden states in the language model, e.g., $h = 768$ for BERT [73]. We feed E into a randomly initialized matrix $W_E \in \mathbb{R}^{\hat{d} \times h}$ to yield an EKS embedding. We repeat M to be of shape $\mathbb{R}^{n \times h}$. By creating all embeddings in the same shape, we facilitate a balanced influence of each input to the model's downstream components. We stack all embeddings to form a matrix $[TEM] \in \mathbb{R}^{n \times 3h}$.

Interaction Layer

We allow the three embeddings to interact using a single-layer BiGRU [151], which yields hidden states $H \in \mathbb{R}^{n \times 6h} = \mathrm{BiGRU}([TEM])$. RNNs, such as LSTMs and GRUs, are commonly used to learn a higher-level representation of a word embedding, especially in state-of-the-art TSC prior to BERT-based models but also recently [151, 208, 213, 401]. We choose an BiGRU over an LSTM because of the smaller number of parameters in BiGRUs, which may in some cases result in better performance [54, 118, 151, 161].

Pooling and Decoding

We employ three common pooling techniques to turn the interacted, sequenced representation H into a single vector [151]. We calculate element-wise (1) mean and (2) maximum over all hidden states H and retrieve the (3) last hidden state h_{n-1}. Then, we stack the three vectors to P, feed P into a fully connected layer FC so that $z = FC(P)$, and calculate $y = \sigma(z)$.

5.3.5   Evaluation

This section describes the experiments we conducted to evaluate our model for target-dependent sentiment classification.

Data and Metrics

In addition to NewsMTSC, we used the three established TSC sets: Twitter, Laptop, and Restaurant. We used metrics established in the TSC literature: macro F1 on all $(F1_m)$ and only the positive and negative classes $(F1_{pn})$, accuracy (A), and average recall (R_a). If not otherwise noted, performances are reported for our primary metric, $F1_m$.

Baselines

We compared our model with TSC methods that yield state-of-the-art results on at least one of the established datasets: SPC-BERT [73]: input is identical to our text input. FC and softmax are calculated on CLS token. TD-BERT [94]: masks hidden states depending on whether they belong to the target mention. LCF-BERT

[400]: similar to TD but additionally weights hidden states depending on their token-based distance to the target mention. We used the improved implementation [394] and enable the dual-LM option, which yields slightly better performance than using only one LM instance [400]. We also planned to test LCFS-BERT [286], but due to technical issues, we were not able to reproduce the authors' results and thus exclude LCFS from our experiments.

Implementation Details

To find for each model the best parameter configuration, we performed an exhaustive grid search. Any number we report is the mean of five experiments that we run per configuration. We randomly split each test set into a dev-set (30%) and the actual test-set (70%). We tested the base version of three LMs: BERT, RoBERTa, and XLNET. For all methods, we tested parameters suggested by their respective authors.[15] We tested all 15 combinations of the following 4 EKS: (1) *SENT* [153], a sentiment dictionary (number of non-mutually exclusive dimensions, 2; domain, customer reviews); (2) *LIWC* [357], a psychometric dictionary (73, multiple); (3) *MPQA* [383], a subjectivity dictionary (3, multiple); and (4) *NRC* [247], dictionary of sentiment and emotion (10, multiple).

Overall Performance

Table 5.8 reports the performances of the models using different LMs and evaluated on both test sets. We found that the best performance was achieved by our model ($F1_m = 83.1$ on *test-rw* compared to 81.8 by the prior state of the art). For all models, performances were improved when using RoBERTa, which is pre-trained on news texts, or XLNET, likely because of its large pre-training corpus. XLNET is not reported in Table 5.8 since its performances were generally similar to those of RoBERTa except for the TD model, where XLNET degrades performance by 5–9pp. Looking at BERT, we found no significant improvement of the proposed model over the prior state of the art. Even if we domain-adapted BERT [300] for 3 epochs on a random sample of 10M English sentences [96], BERT's performance ($F1_m = 81.8$) was lower than RoBERTa. We noticed a performance drop for all models when comparing *test-rw* and *test-mt*. It seems that RoBERTa is better able to resolve in-sentence relations between multiple targets (performance degeneration of only up to −0.6pp.) than BERT (−2.9pp.). We suggest to use RoBERTa for TSC on news, since fine-tuning it was faster than fine-tuning XLNET and RoBERTa achieved similar or better performance than other LMs.

While the proposed model yielded competitive results on previous TSC datasets (Table 5.9), LCF was the top performing model.[16] When comparing the performances across all four datasets, the importance of the consolidation became

---

[15] Epochs $\in \{2, 3, 4\}$; batch size $\in \{8, 16\}$ (due to constrained resources not 32); learning rate $\in \{2 \times 10^{-5}, 3 \times 10^{-5}, 5 \times 10^{-5}\}$; label smoothing regularization (LSR) [353], $\epsilon \in \{0, 0.2\}$; dropout rate, 0.1; $\mathcal{L}_2$ regularization, $\lambda = 1 \times 10^{-5}$; SRD for LCF $\in \{3, 4, 5\}$. We used Adam optimization [181], Xavier uniform initialization [109], and cross-entropy loss.

Table 5.8 Overview of experimental results on NewsMTSC

Model		Test-rw				Test-mt		
		$F1_m$	a $F1_{pn}$		r_a $F1_m$	a $F1_{pn}$	r_a
BERT	SPC	**80.1** 80.7	79.5 79.8		73.7 76.1	71.1 76.0	
	TD	79.4 79.9	78.9 80.0		75.6 79.1	72.0 75.8	
	LCF	79.7 80.9	78.9 79.2		**77.7** 80.5	74.6 79.1	
	GRU	**80.2** 81.1	79.7 80.0		77.3 80.0	74.1 77.9	
RoBERTa	SPC	81.1 82.7	80.5 80.6		79.4 81.6	77.0 79.9	
	TD	81.7 82.5	81.3 81.4		78.4 81.1	75.3 78.2	
	LCF	81.4 82.5	80.8 81.1		81.2 83.8	78.6 81.7	
	GRU	**83.1** 83.8	82.9 83.3		**82.5** 84.6	80.2 81.0	

Table 5.9 Classification performance on previous TSC datasets

Model	Laptop		Restaurant		Twitter	
	$F1_m$	a	$F1_m$	a	$F1_m$	a
SPC	77.4	80.3	78.8	86.0	73.6	75.3
TD	74.4	78.9	78.4	85.1	74.3	77.7
LCF	**79.6**	82.4	**81.7**	87.1	**75.8**	77.3
GRU	79.0	82.1	80.7	86.0	74.6	76.0

apparent, e.g., performance was lowest on the Twitter set, where a simplistic consolidation was employed during the dataset's creation (Sect. 5.3.3.4). The performance differences of individual models when contrasting their use on prior datasets and NewsMTSC highlight the need LCF performed consistently best on prior datasets but worse than the proposed model on NewsMTSC. One reason might be that LCF's weighting approach relies on a static distance parameter, which seems to degrade performance when used on longer texts as in NewsMTSC (Sect. 5.3.3.6). When increasing LCF's window width SRD, we noticed a slight improvement of 1pp. (SRD = 5) but degradation for larger SRD.

Ablation Study

We performed an ablation study to test the impact of four key factors: target mask, EKS, coreferential mentions, and fine-tuning the LM's parameters. We tested all LMs and if not noted otherwise report results for RoBERTa since it generally performed best (Sect. 5.3.5). We report results for *test-mt* (performance influence was similar on either test set, with performances generally being ≈3–5pp. higher on *test-rw*). Overall, we found that our changes to the initial design [151] contributed to an improvement of approximately 1.9pp. The most influential changes were the selected EKS and in part the use of coreferential mentions. Using the target mask input channel without coreferences and LM fine-tuning yielded insignificant improvements of up to 0.3pp. each. We did not test the VADER-based

---

[16] For previous models, Table 5.9 lists results reported by their authors. In our experiments, we found 0.4–1.8pp. lower performance compared to the reported results.

Table 5.10 Classification
influence of exemplary EKS
combinations

Name	$F1_m$	A
no EKS	78.2	81.0
zeros	78.4	81.1
SENT	**80.7**	83.0
LIWC	**80.8**	83.1
MPQA	78.8	80.8
NRC	80.0	82.0
best combination	**81.0**	83.3

Table 5.11 Influence of
target mask and coreferences

Name	BERT	RoBERTa
none	73.1	**78.1**
target mask	73.3	**78.2**
add coref. to mask	**75.6**	**78.1**
add coref. as example	73.0	73.4

sentence classification proposed by Hosseinia, Dragut, and Mukherjee [151] since we expected no improvement by using it for various reasons. For example, VADER uses a dictionary created for a domain other than news and classifies the sentence's overall sentiment and thus is target-independent.

Table 5.10 details the results of exemplary EKS, showing that the best combination (SENT, MPQA, and NRC) yielded an improvement of 2.6pp. compared to not using an EKS (zeros). The single best EKS (LIWC or SENT) each yielded an improvement of 2.4pp. The two EKS "no EKS" and "zeros" represented a model lacking the EKS input channel and an EKS that only yields 0's, respectively.

The use of coreferences had a mixed influence on performance (Table 5.11). While using coreferences had no or even a negative effect in our model for large LMs (RoBERTa and XLNET), it can be beneficial for smaller LMs (BERT) or batch sizes (8). When using the modes "ignore," "add coref. to mask," and "add coref. as example," we ignored coreferences, added them to the target mask, and created an additional example for each, respectively. Mode "none" represents a model that lacks the target mask input channel.

5.3.6   Error Analysis

To understand the limitations of the proposed model, we carried out a manual error analysis by investigating a random sample of 50 incorrectly predicted examples for each of the test sets. For *test-rw*, we found the following potential causes (not mutually exclusive): edge cases with very weak, indirect, or in part subjective sentiment (22%) or where both the predicted and true sentiment can actually be considered correct (10%) and sentiment of given target confused with different target (14%). The latter occurred especially often for long sentences consisting mostly of phrases that indicate the predicted sentiment but concerning a different target, e.g., "By the time he and <u>Mr. Smith</u> (predicted: negative, true: neutral) were trading texts, [...], John was already fired by his boss." Further, sentence's sentiment was unclear due to missing context (10%) and the consolidated answer in NewsMTSC was wrong (10%). In 16%, we found no apparent reason. For *test-mt*, potential causes occurred approximately similarly often as in *test-rw*, except that targets are confused more often (20%).

5.3.7   Future Work

We identify three main areas for future work. The first area is related to the dataset. Instead of consolidating multiple annotators' answers during the dataset creation, we propose to test to integrate the label selection into the model [295]. Integrating the label selection into the machine learning part could improve the classification performance. It could also allow us to include more sentences in the dataset, especially the edge cases that our restrictive consolidation currently discards.

To improve the model design, we propose to design the model specifically for sentences with multiple targets, for example, by classifying multiple targets in a sentence simultaneously. While we early tested various such designs, we did not report them due to their comparably poor performances. Further work in this direction should perhaps also focus on devising specialized loss functions that set multiple targets and their polarity into relation. Lastly, one can improve various technical details of the proposed model, e.g., by testing other interaction layers, such as LSTMs, or using layer-specific learning rates in the overall model, which can increase performance [347].

5.3.8   Conclusion

In this section, we presented NewsMTSC, a dataset for target-dependent sentiment classification (TSC) on news articles consisting of 11.3k manually annotated examples. Compared to prior TSC datasets, the dataset is different in key factors, such as that its texts are on average 50% longer, sentiment is expressed explicitly only rarely, and there is a separate test set for sentences containing multiple targets. In part as a consequence of these differences, state-of-the-art TSC models yielded non-optimal performances in our evaluation.

We proposed a model that uses a bidirectional GRU on top of a language model (LM) and other embeddings, instead of masking or weighting mechanisms as employed by the prior state of the art. We found that the proposed model achieved superior performances on NewsMTSC and was competitive on prior TSC datasets. RoBERTa yielded better results compared to using BERT, because RoBERTa is pre-trained on news and we found it can better resolve in-sentence relations of targets, i.e., RoBERTa can better distinguish the individual sentiments if multiple targets are present in a sentence.

In the context of the PFA approach, TSC represents a method which we propose to use in the target concept analysis component. Conceptually, the TSC method is simpler compared to the fine-grained framing effect classification proposed in Sect. 5.2. However, at the same time, the TSC method represents a pragmatic alternative to imitating part of the manual frame analysis as conducted in social science research on media bias. Due to its simplicity, the TSC method achieves strongly higher classification performance than the approach for frame property identification.

We provide the dataset and code to reproduce our experiments at https://github.com/fhamborg/NewsMTSC.

5.4 Summary of the Chapter

This chapter presented *frame analysis* as the second and last analysis component of person-oriented framing analysis (PFA). The component aims to identify how persons are portrayed in the given news articles, both at the article and sentence levels. This task is difficult for various reasons, such as news articles rather implicitly or indirectly frame persons, for example, by describing actions performed by a person. In sum, reliably inferring how news articles and sentences portray persons is much more complex compared to prior work in related fields. For example, target-dependent sentiment classification (TSC) is concerned with inferring a sentence's sentiment toward a target concept. TSC methods achieve high classification performance, but only on domains where authors explicitly state their attitude toward the targets, such as product reviews or Twitter posts. Because of this difficulty and the other issues highlighted in the chapter, prior approaches concerning frame analysis yield inconclusive or superficial results or require exacting manual effort. Thus, automatically and reliably identifying how persons are portrayed in news articles is essential for the success of PFA. We explored two approaches to enable target concept analysis: event extraction and coreference resolution.

Our first, exploratory approach identifies how a person is portrayed using so-called frame properties (Sect. 5.2). Frame properties are categories that represent predefined, topic-independent effects of political framing. As such, the approach aims to resemble how media bias is analyzed in social science research while avoiding the topic-specific and analysis question-specific frames used there. Early during our research on this approach, we conducted a short qualitative evaluation and found inconclusive results. The inconclusive results and the at that point already very high annotation cost are difficulties common among prior automated approaches that aim to resemble frame analyses (Sect. 5.2.5).

We took these issues as a motivation to explore a more pragmatic route to our frame analysis component. Specifically, we devised a dataset and deep learning model for *target-dependent sentiment classification (TSC)* in news articles (Sect. 5.3). Similar to the frame properties approach outlined previously, the TSC method aims to identify how a person is portrayed using categories representing predefined, topic-independent effects of political framing. In contrast to frame properties, TSC uses only a single dimension as a fundamental effect of framing: polarity, i.e., whether a person is portrayed positively, negatively, or neither. This way, we avoid the infeasibly high annotation cost and ambiguity of analyzing frames or frame derivatives while still capturing an essential framing effect. In contrast to any prior work, our method is the first to reliably classify sentiment in news articles ($F1_m = 83.1$) despite the high level of interpretation required.

In the evaluation described in Chap. 6, we will investigate whether analyzing only a single framing effect dimension, i.e., sentiment polarity, instead of fine-grained framing effects, such as the frame properties, suffices to identify meaningful person-oriented frames.

Chapter 6
Prototype

Abstract This chapter demonstrates the effectiveness of person-oriented framing analysis (PFA) by implementing and evaluating a prototypical system for bias identification and communication. In the single-blind setting of the evaluation, only the PFA prototype consistently, significantly, and most strongly increased respondents' bias-awareness, i.e., respondents' motivation and capabilities to contrast news coverage. The study results and a qualitative analysis indicate that a reason for the improved bias-awareness is that the frames identified by PFA are meaningful and indeed present in person-centric news coverage. In contrast, and confirming the findings of Chap. 2, the most effective baselines only facilitate the visibility of potential frames.

6.1 Introduction

As stated in Chap. 1, empowering newsreaders in recognizing biases in political coverage is crucial since slanted news coverage can decisively impact public opinion and societal decisions, such as in elections. Fitting means for more balanced interaction with news media include practicing media literacy. However, while such non-technical means can be highly effective, they require high effort, such as for researching an event's articles and contrasting their coverage. The high effort may represent an insurmountable barrier, preventing critical assessment in daily news consumption. Automated approaches to effortlessly identify and expose potential biases can complement manual media literacy techniques or even enable them in the first place during daily news consumption (Chap. 3).

This chapter introduces and evaluates *Newsalyze*, our prototype system to reveal biases in news articles by employing person-oriented framing analysis (PFA). While the previous chapters devised methods for PFA and then evaluated their technical performance, this chapter employs a large-scale user study to evaluate the practical effectiveness of the PFA approach in revealing biases. Our goal is to encourage non-expert news consumers to contrast how news articles report on individual events and investigate if our prototype supports its users in doing so.

© The Author(s) 2023
F. Hamborg, *Revealing Media Bias in News Articles*,
https://doi.org/10.1007/978-3-031-17693-7_6

The remainder of the chapter is structured as follows. Section 6.2 summarizes the most related findings of the literature review described in Chap. 2. Section 6.3 introduces our prototype system Newsalyze, which implements PFA by integrating target concept analysis and frame analysis. Section 6.4 introduces layouts and components to build modular visualizations to reveal biases. Section 6.5 presents the study design to evaluate our prototype in a setting that resembles real-world news consumption. In Sect. 6.6, we use two pre-studies to confirm and refine the design and visualizations. Section 6.7 presents the results of the study, and Sect. 6.8 discusses the limitations of both the prototype and the study to derive future research ideas. Lastly, Sect. 6.9 summarizes the main findings of our approach, and Sect. 6.10 concludes this chapter by discussing these findings in the context of this doctoral thesis.

We publish the survey materials, including questionnaires, news articles, visualizations, and anonymized respondents' data at https://doi.org/10.5281/zenodo.4704891.

The source code of the Newsalyze prototype is available at https://github.com/fhamborg/newsalyze-backend/.

6.2   Background

This section briefly defines terms that are relevant for our study (see Sect. 6.2.1) and summarizes prior work relevant for the tasks of bias identification and bias communication (see Sect. 6.2.2). More in-depth information concerning the reviewed approaches can be found in Chap. 2.

6.2.1   Definitions

We use our definition of media bias as introduced in Sect. 3.2. Specifically, we define *bias* as the effect of framing, i.e., the promotion of "a particular problem definition, causal interpretation, moral evaluation, and/or treatment recommendation" [79] that arises from one or more of the bias forms defined by the news production process.

We define *bias-awareness* generally as an effect of bias communication. In practical terms, we define bias-awareness in this chapter as an individual's motivation and ability to relate and contrast perspectives present in news coverage, both to another [276] and also to the individual's views [104].

6.2.2 *Approaches*

Following the thesis's research objective (Sect. 3.3.3), we briefly summarize approaches for the analysis and communication of media bias (also called bias diagnosis, measurement, and mitigation [276]). In this chapter, we exclude other means to tackle media bias, such as bias prevention during the production of news because they cannot practically tackle media bias (see our discussion of the solution space in Sect. 3.3).

Our literature review on prior work concerned with the analysis or communication of media bias reveals that bias-sensitive visualizations can effectively increase news consumers' bias-awareness. Thus, such approaches may in principle support news consumers in making more informed choices [22]. However, we also find that the reviewed approaches suffer from one or more of the following shortcomings.[1]

High Cost and Lack of Recency Content analyses and frame analyses are among the most effective bias analysis tools. Decade-long research in the social sciences has proven them effective and reliable, e.g., to capture also subtle yet powerful biases (cf. [79]). However, because researchers need to conduct these analyses mostly manually, the analyses do not scale with the vast amount of news (Sect. 2.2.4). In turn, such studies are always conducted for few topics in the past and do not deliver insights for the current day [228, 267]; this would, however, be an effective means to support readers in critically assessing the news during daily news consumption (see Sect. 3.3.3).

Superficial Results Many automated approaches for bias identification suffer from superficial results, especially when compared to the results of analyses as conducted in the social sciences (Sect. 2.5). Reasons include that the approaches treat media bias as a rather vaguely or broadly defined concept, e.g., "differences of [news] coverage" [278], "diverse opinions" [251], or "topic diversity" [252], and neglect social science bias models (see Chap. 2). Further, especially early approaches [252, 276] suffer from poor performance since word-, dictionary-, or rule-based methods as commonly employed in traditional machine learning fail to capture the "meaning between the lines" [126]. To improve performance, some approaches employ crowdsourcing [8, 277, 332], e.g., to gain bias ratings. Crowdsourcing can be an effective means to gather labeled data. However, such data is problematic if not carefully reviewed for biases [154], e.g., if users are not a representative sample or already biased through earlier exposure to systematically biased news coverage. Other approaches approximate biases by grouping news articles according to their news outlets' respective political orientation [8]. Recent methods that employ deep learning or word embeddings can yield more substantial results, e.g., they identify framing categories that reflect meaningful patterns in the analyzed news texts. However, the creation of large-scale datasets required for their training is very costly

---

[1] An in-depth discussion of the following and related approaches can be found in Chap. 2.

(Sect. 5.1), and semi-automated approaches require careful manual revision of the automatically identified bias categories [193].

Inconsistency The design of some approaches only facilitates the visibility of biases that might be in the data rather than identifying meaningful biases present in the data. Reasons for this inconsistency partially overlap with the previously mentioned reasons for automated approaches' superficial results. Additional reasons include that approaches do not analyze the articles' content to determine their biases but approximate potential biases using metadata, such as the political orientation of the articles' outlets [8]. Others analyze the content but use only shallow or non-representative features, e.g., they analyze only headlines but not the remainder of articles [187].

Besides, many of the previously mentioned approaches are expert systems and thus not suitable for daily news consumption. While there are some easy-to-use systems and visualizations, especially outside the academic context, they suffer either from the previously mentioned shortcoming concerning bias identification [8] or are entirely bias-agnostic. For example, Fig. 6.1 depicts the bias-agnostic news overview provided by Google News. The main part in the center shows a list of current news events, where for each event, multiple articles reporting on it are shown. No information is available concerning how these articles are selected. However, we find that they are selected to represent the event and fit the users' preferences, e.g., are from their favorite news outlets or are geographically close.

Besides its superficial bias analysis approach, AllSides entails an easy-to-use visualization for bias communication [8], which is intended to reveal biases present in current event coverage quickly. Figure 6.2 depicts the news overview provided by AllSides. Like Google News overview, a list shows current events and, for each event, multiple articles reporting on the event. In contrast to Google News and other popular news aggregators, AllSides aims to inform users about the different perspectives present in the event coverage. Therefore, AllSides shows at the top one event-representative article and below three articles, each representing a left-wing, center, and right-wing news outlet.

In sum, most prior studies confirm the effectiveness and benefits of communicating biases to news consumers. However, prior work suffers from various shortcomings, such as requiring manual analyses, yielding superficial results, only facilitating the visibility of media bias that *might* be in the data, or requiring training before their use.

6.3   System Description

This section introduces our prototype *Newsalyze*. The prototype integrates the methods devised previously in this thesis to implement the person-oriented framing analysis (PFA) approach.

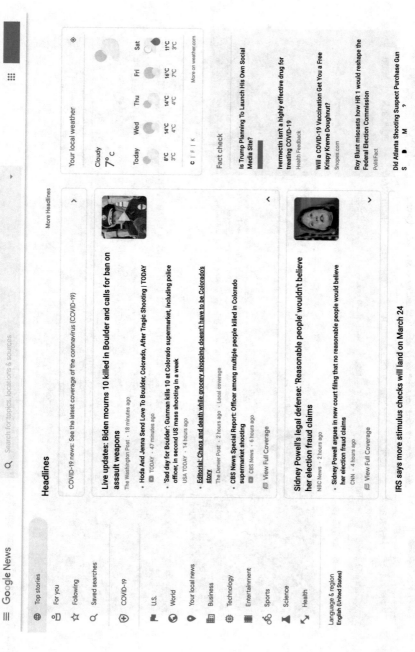

Fig. 6.1 Screenshot of the bias-agnostic news overview provided by Google News

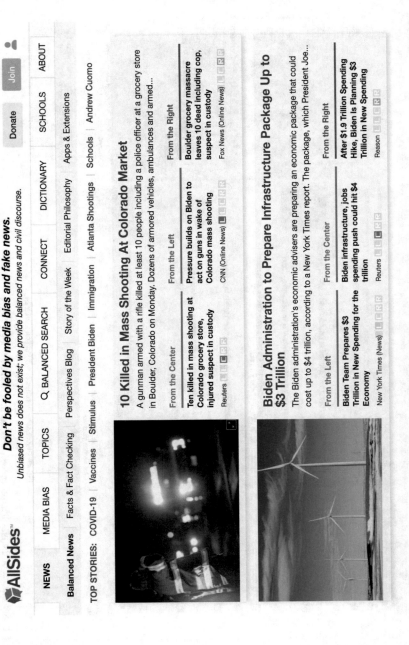

Fig. 6.2 Screenshot of the bias-sensitive news overview provided by AllSides

Given a set of news articles reporting on the same political event, our system seeks to find and visualize groups of articles that similarly frame the persons involved in the event using three phases: article gathering, bias analysis, and bias communication. This section summarizes our previous research concerning article gathering (Sect. 3.5) and bias analysis using PFA (Sect. 3.4). Section 6.4 then introduces our novel visualizations for bias communication.

For article gathering, we integrate our crawler and extractor specifically tailored for news articles (Sect. 3.5). Users provide the system with a set of URLs linking to news articles reporting on the same event to be analyzed by Newsalyze. The news crawler then extracts the required information from the articles' web pages, i.e., title, lead paragraph, and main text. Alternatively, users can directly provide news articles to the system, e.g., by providing JSON files containing the previously mentioned information.

The bias identification using PFA consists of the three tasks depicted in Fig. 3.1 in Sect. 3.4. First, we perform NLP *preprocessing* as described in Sect. 4.3.3.1. We use our *split preprocessing* (Sect. 4.3.3.1) since it yields better coreference resolution performance than the standard preprocessing (Sect. 4.3.4.3).

In the following, we describe the subsequent tasks of PFA, i.e., target concept analysis and frame analysis.

Target concept analysis finds and resolves persons mentioned across the topic's articles, including highly event-specific coreferences as they frequently occur in person-targeting bias forms. As highlighted in Chap. 1 and Sect. 2.3.4, especially in the presence of bias by word choice and labeling, persons' mentions may be coreferential only in the coverage on a specific event, but otherwise they may be not be coreferential or even opposing in meaning, such as "regime" and "government." To resolve such mentions, we use the method for context-driven cross-document coreference as described in Sect. 4.3. Specifically, we use the method using the first two sieves since they suffice to achieve the highest performance on individual persons (see concept type *Actor* in Sect. 4.3.4.3). The output of target concept analysis is the set of persons involved in the news coverage of the event and, for each person, all the person's mentions across all news articles.

Frame analysis determines how the news articles portray the persons involved in the event and then finds groups of articles that similarly portray these persons. This task centers around our concept of person-targeting framing, which resemble (political) framing as defined by Entman [79], where a frame "promotes a particular problem definition, causal interpretation, moral evaluation, and/or treatment recommendation" [79]. Our person-oriented frames resemble these political frames, but are somewhat exploratory, e.g., implicitly defined and loosely structured.[2] As we discuss in Sect. 3.3.2, identifying frames would approximate content analyses, the standard tool used in the social sciences to analyze media bias (Sect. 2.2.4). However, doing so would require infeasible effort since researchers in the social sciences typically create frames for a specific research question [45, 46, 79].

---

[2] More information on both concepts is described in Sect. 3.3.2.

PFA, however, is meant to analyze media bias on any coverage reporting on policy issues. Thus, we seek to determine a fundamental bias effect resulting from framing: polarity of individual persons, which we identify for each person mention (on the sentence level) and aggregate to article level. To achieve state-of-the-art performance in target-dependent sentiment classification (TSC) on news articles, we use our RoBERTa-based [214] neural model trained on our dataset (Sect. 5.3).

The last step of frame analysis is to determine groups of articles that frame the event similarly, i.e., the persons involved in the event. We call the resulting groups *framing groups*. By definition, all articles of one framing group share the same person-oriented frame, i.e., they represent one perspective present in the event coverage. We propose two methods for grouping. (1) *Grouping-MFA*, a simple, polarity-based method, first determines the single person that occurs most frequently across all articles, called *most frequent actor (MFA)*. Then, the method assigns each article to one of three groups, depending on whether the article mentions the MFA mostly positively, ambivalently, or negatively. (2) *Grouping-ALL* considers the polarity of all persons instead of only the MFA. Specifically, the method uses k-means with $k = 3$ on a set of vectors where each vector a represents a single news article:

$$a = \begin{pmatrix} s_0 \\ \vdots \\ s_i \end{pmatrix}, \tag{6.1}$$

where $i \in (0, \ldots, |P| - 1)$, P the set of all persons, a person's sentiment polarity s_i in a is

$$s_i = \sum_{m \in M} \frac{w(m)s(m)}{m_{\max,a}}, \tag{6.2}$$

where m is each mention of all the person's mention in a, $w(m)$ is a weight depending on the position of the mention (mentions in the beginning of an article are considered more important [53]), and $s(m)$ yields the polarity score of m (1 for positive, -1 for negative, 0 else). To consider the individual persons' frequency in an article for clustering, we normalize by $m_{\max,a}$, which is the number of mentions of the most frequent person in a.

In addition to grouping, we calculate each article's relevance to the event and the article's group using simple word embedding scoring.

6.4   Visualizations

Our visualizations aim to aid in typical online news consumption, i.e., an *overview* enables users to first get a synopsis of news events and articles (Sect. 6.4.1) and an *article view* shows an individual news article (Sect. 6.4.2). We seek to devise visualizations that (1) are easy to understand, i.e., usable by non-experts without prior training, and that (2) reveal biases (see our research objective described in Sect. 3.3.3). To measure the effectiveness not only of our visualizations but also their constituents, we design them so that individual visual features can be altered or exchanged. Later, in our conjoint-based evaluation [122], we can measure the effects of each constituent, i.e., the individual visual clues. In the following, a "conjoint profile" refers to a specific combination of all visual clues. For example, one specific conjoint profile of the news overview would show certain visual clues with specific settings while not showing other visual clues. The conjoint design will be described in detail in Sect. 6.5.2.

To more precisely measure the change in bias-awareness concerning only the textual content as required by our research objective (Sect. 3.3.3), we apply changes compared to typical news consumption. For example, the visualizations show the texts of articles (and information about biases in the texts) but no other content, e.g., no photos or outlet names. Further, in our study, the overview shows only a single event instead of multiple.

6.4.1   Overview

The overview aims to enable users to get a synopsis of a news event quickly. We devise a modular, bias-sensitive visualization layout, which we use to implement and test specific visualizations. The comparative layout aims to support users in quickly understanding the frames present in coverage on the event. The layout is vertically divided into three parts, denoted as parts *a*, *b*, and *c* in Fig. 6.3. The event's *main article* (part a) shows the event's most representative article. The *comparative groups* part (b) shows up to three perspectives present in event coverage by showcasing each perspective's most representative article. It is designed to encourage users to contrast the articles and critically assess their content. When employing PFA, these perspectives are person-oriented frames. Since we also test baselines representing the state of the art, we generally refer to perspectives in this section. To determine the framing groups, the system uses one of the grouping methods described in Sect. 6.3, i.e., Grouping-MFA and Grouping-ALL. Finally, a list shows the headlines of *further articles* reporting on the event (part c).

All visualizations show brief explanations for all features that users may not be familiar with. For example, the overview contains a brief explanation informing users about what the comparative groups represent and how they were derived (see "1" in Fig. 6.4 for the specific variant of Grouping-MFA and "1" in Fig. 6.5 for the generic variant used by any grouping).

In each overview, two types of visual clues conveying bias information can be enabled and altered depending on the conjoint profile (see Sect. 6.5.2). First, zero or more headline tags are shown next to each article's headline. They indicate the political orientation of the article's outlet (*PolSides tags*; see "2" in Fig. 6.4), the

Tags shown near a headline denote the political orientation (as self-identified by the publisher) of its article (`left` `center` `right`) and how the article possibly portrays (determined automatically) the topic's main person **President Donald Trump** (`possibly contra` `possibly ambivalent` `possibly pro`).

A

Trump, Congress Reach Agreement On 2-Year Budget Deal `center` `possibly pro`

President Trump announced an agreement on a two-year budget deal and debt-ceiling increase. The deal would raise the debt ceiling past the 2020 elections and set $1.3 trillion for defense and domestic spending over the next two years. [...]

Each article is assigned to either of the following groups depending on how the article portrays the main person. Below, you see for each group its most representative article. To determine how an article reports on the main person, we automatically classify the sentiment of all mentions of that person. In this topic, the main person is:

President Donald Trump

B

Possibly pro	Possibly ambivalent	Possibly contra
Trump, Congress Clinch Debt-Limit Deal After Tense Negotiations `center`	**Donald Trump, congressional Democrats reach two-year budget deal, avoid crisis on debt ceiling** `center`	**Donald Trump and the G.O.P. Confirm Their Fiscal Conservatism Was a Sham** `left`
President Donald Trump announced a bipartisan deal to suspend the U.S. debt ceiling and boost spending levels for two years, capping weeks of frenzied negotiations that avert the risk of a damaging payments default. [...]	WASHINGTON — The White House and congressional leaders have reached a new budget deal that calls for raising federal spending levels and lifting the debt ceiling for two years, potentially averting what could have been another nasty partisan battle this fall. [...]	This week, the Republican Party, with its eyes on November, 2020, and with encouragement from Trump, said to heck with the deficit and the debt. [...]

C

Further articles

▼ **White House, congressional Democrats agree on debt ceiling hike** `right` `possibly pro`

The White House and congressional Democrats agreed Monday on a two-year budget deal that settles on a new debt ceiling and would likely eliminate the risk of a government shutdown this fall. [...]

▶ **Trump announces 'real compromise' on budget deal, as fiscal hawks and some Dems cry foul** `right` `possibly ambivalent`

▶ **Trump Announces Deal On Debt Limit, Spending Caps** `left` `possibly ambivalent`

Fig. 6.3 Newsalyze's overview consists of three parts: the *main article* (part (**a**)) represents a single; the *comparative groups* (part (**b**)) showcase up to three perspectives in the event coverage; and *further articles* (part (**c**)) are a list of further articles reporting on the event

article's overall polarity regarding the MFA due to Grouping-MFA (*MFAP tags*; see "3"), and the article's group according to its polarity regarding all persons due to Grouping-ALL (*ALLP tags*), respectively.

Second, labels and explanations in the visualization are either *generic* or *specific*. The specific variants explain how the grouping was specifically performed

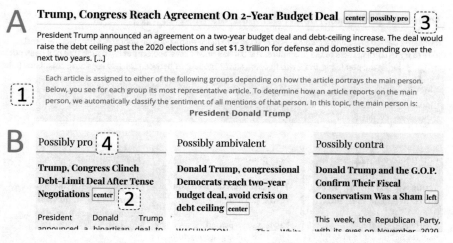

Fig. 6.4 Excerpt of the news overview showing three perspectives of news coverage on a debt ceiling event

(see "1" in Fig. 6.4) and provide specific group labels (see "4" in Fig. 6.4). In contrast, all visualizations employing the generic variant use the same universal explanation, e.g., only mentioning that our system automatically determined the three perspectives (see "1" in Fig. 6.5), and use the same generic coloring and labels, e.g., "Perspective 1" as shown close to "3" and "4" in Fig. 6.5.

6.4.2 Article View

The article view visualizes a single news article. It thus represents the second step in typical news consumption, i.e., after getting an overview of current events, users subsequently may want to read individual articles of interest. The layout of the view is vertically divided into three parts, denoted as parts *a*, *b*, and *c* in Fig. 6.6. The *bias information* part (a) at the top contains various visual elements that aim to inform newsreaders about the bias and positioning of the current article. The *main* part (b) shows the given article's headline, lead paragraph, and main text. Lastly, a list shows the headlines of *further articles* reporting on the event (part c). Within these three parts, various visual clues to communicate bias information are enabled, disabled, or altered depending on the conjoint profile. We describe them in the following.

The *bias information* part ("a" in Fig. 6.6) contains up to three visual clues to inform about potential slants of the current article. Specifically, the *polarity context bar* aims to enable users to quickly understand the overall slant concerning the event's MFA of the current and other articles. The 1D scatter plot depicted in Fig. 6.7 represents each article as a circle. The polarity context bar places each circle depending on its article's overall polarity regarding the MFA. To quickly assess how

Tennessee passes abortion restriction bill Perspective 2 3

Tennessee passed sweeping legislation Friday seeking to place restrictions on abortion, becoming the latest state to try to curtail access to the procedure. The bill was one of several pieces of legislation that had been put on pause during the pandemic, but it was boosted after negotiations between the two legislative chambers. [...]

1

When reviewing media coverage on the given news topic, we identified three main perspectives. We determined the main perspective of each article. Below, you see for each perspective its most representative article.

Perspective 1 4	Perspective 2	Perspective 3
Tennessee advances 6-week abortion ban, lawsuit filed	**Tennessee passes abortion restriction bill**	**Tennessee lawmakers pass 'heartbeat' abortion bill banning procedure after six weeks**
NASHVILLE, Tenn. — Amid nationwide unrest and a global pandemic that wrecked the state budget, Tennessee lawmakers advanced one of the strictest abortion bans in the country as most Tennesseans were asleep Friday and largely unaware the GOP-dominant General Assembly had taken up the controversial proposal. [...]	Tennessee passed sweeping legislation Friday seeking to place restrictions on abortion, becoming the latest state to try to curtail access to the procedure. The bill was one of several pieces of legislation that had been put on pause during the pandemic, but it was boosted after negotiations between the two legislative chambers. [...]	Tennessee lawmakers passed one of the tightest abortion restrictions in the country on Friday, banning the procedure once a fetal heartbeat is detected at around six weeks, which is often before a woman realizes she is pregnant. [...]

Further articles

▼ **Tennessee lawmakers pass fetal heartbeat abortion bill backed by governor** Perspective 1

Washington Tennessee lawmakers have passed a bill backed by the state's Republican governor Bill Lee that would ban abortions after a fetal heartbeat is detected. Early Friday morning, the Tennessee Senate approved the bill, after the House had passed the legislation earlier. Republicans control both chambers. [...]

▶ **Tennessee Legislature Passes Fetal Heartbeat Bill** Perspective 2

▶ **Tennessee passes 6-week abortion ban during last-second budget negotiation** Perspective 1

Fig. 6.5 Shown is a news overview where the specific explanations, e.g., how the grouping was performed, and labels, e.g., headline tags, are replaced with generic variants. The added labels ("1," "3," and "4") refer to the same as depicted in Fig. 6.4

the current article's polarity compares to the other articles' slants, the current article is highlighted using a bold circle (see "1" in Fig. 6.7). Users can interactively, i.e., by hovering their cursor over the circles, view individual articles' headlines (see "2").

Also within the bias information part, *bias indicators* show the article's framing group, analogously to the headline tags, i.e., the outlet's political orientation (PolSides) and how the article reports on the MFA, called *MFAP* (as identified by Grouping-MFA; see Sect. 6.3), or all persons, called *ALLP* (as identified by

A Information about the article

This article has the following **political orientation**:

| left | center | right |

How articles that report on the topic portray (determined automatically) the topic's main person **Prime Minister Scott Morrison**:

Possibly contra ├─────────────○─────────────┤ Possibly pro

this article

B Article

Color coding indicates a sentence's possible stance toward the highlighted person (pro (green), ambivalent (yellow), or contra (red)).

Climate change or poor policy? As Australia's wildfires see some relief, blame game ascends

The sky finally cracked open atop much of the fire grounds in NSW, Australia this week – the welcome deluge slashing the number of burning fires to below 100 for the first time in more than a month.

But the challenge is far from over as the finger-pointing game ignites.

" There are many reasons for the fires starting," Paul Baxter, Commissioner of Fire and Rescue NSW and President and Board Chair of the Australasian Fire and Emergency Services Authorities Council (AFAC), told Fox News. " Some have been natural – lightning – and some have been caused by human, both maliciousness and carelessness.

" Climate change has brought bountiful rains throughout most of the past two decades, which have suppressed wildfires and allowed for more vegetation growth. That is a good thing, " said James Taylor, Director of the Arthur B. Robinson Center for Climate and Environmental Policy at The Heartland Institute. " However, Australian government policies that discourage or prohibit prescribed burns and other proactive land management have meant that when we finally have a dry year, there is more fuel for the fires.

" All considered, climate change is suppressing Australian wildfires, but poor government policies, ironically supported by climate and environmental activists, are making the present wildfires worse," Taylor conjectured.

Conservative Prime Minister Scott Morrison has pledged to "address issues around hazard reduction for national parks, dealing with land clearing laws, zoning laws and planning laws around people's properties and where they can be built."

Morrison has furthermore denied accusations that he has failed to acknowledge climate change during his tenure in office. But he also has remained steadfast in protecting the coal industry and the hundreds of thousands of families who rely on it to make ends meet.

Arson, too, has played its part.

Police in New South Wales this week announced that, since the beginning of November, two dozen people have been charged with intentionally lighting fires, while 53 others have been accused of failing to comply with total fire ban regulations.

Scientists insist that climate change has acerbated already ripe conditions for calamitous fires, while critics have contended that such devastation is nothing new to the Australian landscape.

"This is not Australia's hottest year. Bushfires are a common occurrence there, going back a long time," argued Steve Milloy, publisher of JunkScience.com. "The 1939 bushfires, for example, were much worse. The most important reason for the bushfires is poor land management practices. "

But for those most impacted by the combustion – some described shooting their animals, their faces awash with tears – skepticism remains that it will never really be over.

C Further articles

Tags shown near a headline denote the political orientation (as self-identified by the publisher) of its article (left center right).

- **Australians urged to flee as huge wildfires revitalize** left
- **Australia's Wildfires Spark Disinformation Battle As They Take A Tragic Toll** center
- **'Apocalypse': 500 million animals estimated dead as Australian wildfires rage** right

Fig. 6.6 Newsalyze's article view consists of three parts: the *bias information* part (**a**) shows bias-related information concerning the given article; the *main* part (**b**) contains the given article; and *further articles* (part (**c**)) are a list of further articles reporting on the event

Fig. 6.7 Polarity context bar showing the current and other articles' polarity regarding the MFA and a tooltip of the headline of a hovered article

Fig. 6.8 PolSides bias indicator showing the current article's political orientation as identified by its outlet

Overall, this article is regarding the topic's main person **Prime Minister Scott Morrison** possibly:

contra   ambivalent   pro

Fig. 6.9 MFAP bias indicator showing the current article's overall polarity concerning the MFA (here, the MFA "Prime Minister Scott Morrison" is shown ambivalently)

Grouping-ALL). In contrast to the headline tags, which are shown besides all headlines, each indicator is a component that prominently shows the framing group of only the current article. Depending on the conjoint profile (identical to headline tags), individual indicators are shown or disabled. Figure 6.8 depicts the PolSides bias indicator; Fig. 6.9 depicts the MFAP bias indicator.

Within the *main* part of the article view ("b" in Fig. 6.6), *in-text polarity highlights* aim to enable users to quickly comprehend how the individual sentences of the current news article portray the mentioned persons. To achieve this, we visually mark mentions of individual persons within the news article's text. We test the effectiveness of the following modes: *single-color* (visually marking a person mention using a neutral color, i.e., gray, if the respective sentence mentions the person positively or negatively), *two-color* (using green and red colors for positive and negative mentions, respectively), *three-color* (same as two-color and additionally showing neutral polarity as gray), and *disabled* (no highlights are shown). For example, in the sentence "The Mueller report was tough on Trump," the person mention "Trump" has negative polarity and would be highlighted red in the two- and three-color modes.

Within the *further articles* list ("C" in Fig. 6.6), *headline tags* are shown and have identical purpose and function as when shown in the overview (see Sect. 6.4.1).

6.5 Study Design

In this section, we present our study design to measure the effectiveness of our system. In contrast to prior bias studies, our design allows us to pinpoint the effectiveness to individual components. Section 6.7 presents our results, and Sect. 6.8 discusses the study's limitations to derive future work ideas. All survey data, including questionnaires and anonymized respondents' information, is available freely (see Sect. 6.1).

6.5.1 Objectives and Questions

We base our study design on our definition of bias-awareness (Sect. 6.2) as the primary metric to investigate the effectiveness of an analyzed means to "reveal biases" as requested by our research question (Sect. 1.3). In particular, the definition of bias-awareness highlights the need to contrast perspectives in the news as an effective means to become aware of biases, which in turn are defined as just these perspectives.

We focus in our study on the overview visualizations, with their comparative groups being the primary means to enable contrasting perspectives and thus reveal biases.

Q1: How does a bias-sensitive, easy-to-understand news overview improve bias-awareness in non-expert news consumers?

Secondarily, we seek explore how bias-awareness can be affected by revealing biases in individual articles and by the respondents themselves.

Q2: How does a bias-sensitive, easy-to-understand article view improve bias-awareness in non-expert news consumers?
Q3: How do demographic factors of news consumers affect their bias-awareness?

6.5.2 Methodology

We propose to use a conjoint design [218], which is especially suitable for estimating the effect of individual components. Traditional survey experiments are limited to only identifying the "catch-all effect" [122] due to confounding of the treatment components. In contrast, conjoint experiments identify "component-specific causal effects by randomly manipulating multiple attributes of alternatives simultaneously" [122]. In a conjoint design, respondents are asked to rate so-called profiles, which consist of multiple *attributes*. In our study, such attributes are, for example, the overview, which topic it shows (or which article the article view shows), and if or which tags or in-text highlights are shown.

Conjoint experiments rest on three core assumptions: (1) stability and no carry-over effects, (2) no profile-order effects, and (3) randomization of the profiles [122]. In our evaluation, (2) holds by design for all tasks except for the forced-choice question (see workflow step 6 in Sect. 6.5.6). We briefly describe our means to ensure (1) and (3) in the following.[3]

To ensure (1), i.e., the absence of carry-over effects from one task set to another, we applied during the study the diagnostics proposed by Hainmueller, Hopkins, and Yamamoto [122]. We refer to a task set as all tasks shown to a respondent for a single topic, e.g., in our main study, we show respondents for each topic one overview and subsequently two article views. We then calculated if there are meaningful differences across the task sets by building a sub-group for each task set. We found weak carry-over effects when comparing the individual attributes' effects (using our main overview question across the task sets) and when testing the effect of the task set's order (for all overview questions combined (Est. $= 3.28\%$, $p = 0.018$), i.e., respondents were on average more bias-aware in the second task set). Further, in our main study, when sub-grouping for the task set, the other attributes' effects differed. However, this is not necessarily problematic. A learning effect is expected and desirable in bias communication. Since we randomized the attributes within each task set, we can include the task sets in the analysis and thus measure the effects, regardless of the task set.

We ensure (3) by randomly choosing the attributes independently of another and for each respondent. To confirm the randomization was successful in our experiments, we employed a Shapiro-Wilk test during the study [296].

Since in each of our experiments the three assumptions hold, our design allows for an estimation of the relative influence of each component on the bias-awareness, which is called average marginal component effects (AMCEs) [122]. An AMCE represents the effect of an attribute level, e.g., the two-color mode (attribute level) of our in-text highlights (attribute), compared to a pre-selected baseline of that attribute, e.g., not showing any in-text highlights. In simplified terms, the concept of AMCEs is to create two subsets, one for the current attribute level and one for the attribute baseline. Then, the respondents' answers, e.g., to our questions in the post-article questionnaire, are averaged in each subset. Lastly, by comparing the averaged answers of both sets, the AMCE represents the increase or decrease of an attribute's specific level compared to the attribute's baseline.

In our questionnaires, we employ discrete choice (DC) as well as rating questions (see Sect. 6.5.6) to measure bias-awareness on a behavioral as well as attitudinal

---

[3] We describe the means here for improved readability, even though we executed these means for each of our pre-studies and the main study.

level [303]. DC questions are widely used within the conjoint design and found to have high external validity in mimicking real-world behavior [121]. Additionally, DC questions elicit behavior, i.e., which news article respondents prefer to read or rely on for decision-making [287]. In contrast, rating questions capture attitudes and personal viewpoints better [380].

6.5.3   Data

We select four news topics with varying degrees of expected biases among the news articles reporting on the topic. To approximate the degree of bias, we use the topics' expected polarization. Specifically, we select three topics expected to be highly polarizing for US news readers: gun control (Orlando shooting in 2016), debt ceiling (discussions in July 2019), and abortion rights (Tennessee abortion ban in June 2020). To better approximate regular news consumption, where consumers typically are exposed to news coverage on single events, we select a single event for each of these topics (shown in parentheses). We add a fourth event, which we expect to be only mildly polarizing: Australian bush fires, i.e., a foreign event without direct US involvement. As described later, during our pre-studies, we added the abortion topic due to a negative influence of the debt ceiling topic on bias-awareness. In the pre-studies, we could trace this back qualitatively to respondents' critique of the topics being too "boring" and "complicated," which also manifested in lower reading times on average.

For each event, we select ten articles from left-wing, center, and right-wing, online US outlets (political orientation as self-identified by the outlets or from [8]). To ensure high quality, we manually retrieve the articles' content. Before the second pre-study, we shortened all articles so that they were of similar length (300–400 words) to address the high reading times and noise in the responses, a key finding of the first pre-study (see Sect. 6.6). We consistently apply the same shortening procedure to preserve the perspectives of the original articles, for example, by maintaining the relative frequency of person mentions and by discarding only redundant sentences that do not contribute to the overall tone. In all experiments, we remove any non-textual content, such as images, to isolate the effects in the change of bias-awareness due to the text content, our text-centric bias analysis, and visualization.

6.5.4   Setup and Quality

We conduct our experiments as a series of online studies on Amazon Mechanical Turk (MTurk). To participate in any of our studies, crowdworkers have to be located in the USA. To ensure high quality, we further require that participants possess MTurk's "Masters" qualification, i.e., have a history of successfully completed, high-quality work. While we compensate respondents always, our study design includes discarding data of any respondent who fails to meet all quality criteria, including a minimum study duration, and correctly answering questions checking attention and seriousness [12]. Depending on the study's duration, participants receive an assignment compensation that approximates an hourly wage of $10.

6.5.5   Overview Baselines

To answer our primary study question Q1, our study design compares our system and the overview visualization variants with baselines that resemble news aggregators popular among news consumers and an established bias-sensitive news aggregator (cf. [276]).

Plain is an overview variant that resembles popular news aggregators (Fig. 6.10). Using a bias-agnostic design similar to Google News, this baseline shows article headlines and excerpts in a list sorted by the articles' relevance to the event. Section 6.3 describes the relevance calculation. Compared to our bias-sensitive overview design described in Sect. 6.4.1, Plain does not contain the comparative groups (part c in Fig. 6.3) but only the main article and list further articles (parts a and b). Besides each headline, headline tags as described in Sect. 6.4.1 can be shown, depending on the conjoint profile.

PolSides is an overview variant that aims to resemble the bias-sensitive news aggregator AllSides [8]. PolSides yields framing groups by grouping articles depending on their outlets' political orientation (left, center, and right, as self-identified by them or taken from [8]). The visualization uses the same bias-sensitive layout consisting of three vertical parts as we described in Sect. 6.4.1 and show

---

Tennessee passes abortion restriction bill

Tennessee passed sweeping legislation Friday seeking to place restrictions on abortion, becoming the latest state to try to curtail access to the procedure. The bill was one of several pieces of legislation that had been put on pause during the pandemic, but it was boosted after negotiations between the two legislative chambers. [...]

Further articles

▶ **Tennessee Legislature Votes To Ban Abortions After A Heartbeat Can Be Detected**

▶ **Tennessee lawmakers pass fetal heartbeat abortion bill backed by governor**

▼ **Tennessee Legislature Passes Fetal Heartbeat Bill**

Tennessee lawmakers on Friday passed a "heartbeat bill" banning abortion once a fetal heartbeat is detected —a regulation on abortion that has run into legal challenges in several states. The bill makes any abortion performed after a fetal heartbeat is detected a felony. Detection of the fetal heartbeat generally occurs six weeks into a pregnancy, and the bill prohibits abortions based on a fetus's race, sex, or disability. The bill also creates exceptions for situations where the health of the mother is in danger. [...]

▶ **Tennessee legislature passes fetal heartbeat bill, ban on abortions for Down syndrome**

▶ **Tennessee Gov. Bill Lee to Sign Heartbeat Abortion Bill**

Fig. 6.10 The *Plain* news overview uses a list to show articles reporting on a topic

Tennessee passes abortion restriction bill [center]

Tennessee passed sweeping legislation Friday seeking to place restrictions on abortion, becoming the latest state to try to curtail access to the procedure. The bill was one of several pieces of legislation that had been put on pause during the pandemic, but it was boosted after negotiations between the two legislative chambers. [...]

Each article is assigned to either of the following groups depending on its news outlet's political orientation (left, center, right). We use the political orientation as self-identified by the outlet. Below, you see for each group its most representative article.

From the left	From the center	From the right
Tennessee advances 6–week abortion ban, lawsuit filed	**Tennessee passes abortion restriction bill**	**Tennessee Legislature Passes Fetal Heartbeat Bill**
NASHVILLE, Tenn. — Amid nationwide unrest and a global	Tennessee passed sweeping legislation Friday seeking to place	Tennessee lawmakers on Friday passed a "heartbeat bill" banning

Fig. 6.11 The *PolSides* news overview aims to resembles the bias-sensitive news aggregators AllSides, which groups articles depending on their outlets' political orientation

in Fig. 6.3. Figure 6.11 shows an excerpt of the main article and the comparative groups (parts a and b). Conceptually, PolSides employs the left-right dichotomy, a simple yet often effective means to partition the media into distinctive slants, which is also one of the most commonly studied dimensions of bias. Being a well-known and easy-to-interpret concept, we do also expect that users will initially understand PolSides's approach to determine the framing groups. However, this dichotomy is determined only on the outlet level. It thus may incorrectly classify the biases indeed present in a specific event (see Sect. 6.2.2), e.g., articles shown to be of different slants having indeed similar perspectives (and vice versa). We investigate this issue in our study (see Sect. 6.7).

To our knowledge, the baselines exhaustively cover the relevant prior work, particularly concerning the communication of biases in news articles to non-expert news consumers (see Sect. 6.2). While we deem NewsCube another conceptually very relevant approach because of its similar research objective, the visualizations proposed by Park et al. [276] are designed to show only a single article rather than providing a news overview and thus do not allow comparison.

As stated in Sect. 6.4, we are interested in the effects on the bias-awareness due to textual means, i.e., bias forms at the text level. Thus, all visualizations, including the baselines, show the texts of articles and information about biases due to the textual bias forms defined in Sect. 3.3.3 but no other content, e.g., no photos or outlet names.

To understand how *visualizations*, including their layout and explanations, affect bias-awareness compared to the visualized *content*, e.g., the framing groups resulting from our analysis, we introduce two additional baseline concepts. First, we include for most overviews, including the baselines previously mentioned, generic variants (see Sect. 6.4.1). This *single-blind setting* helps to assess how respondents are affected by knowing (such as the popular left-right dichotomy employed by PolSides) or not knowing (such as our novel PFA approach) the employed grouping mechanisms. Second, we test an overview with generic explanations that randomly assign individual news articles to either of the three framing groups.

Concerning our secondary study question Q2, we test two headline tags (PolSides and MFAP) jointly with their respective indicators showing the article's bias classification (PolSides and MFAP), each as described in Sect. 6.4.2. For example, if PolSides headline tags (showing the political orientation of each article in the list of further articles) are enabled, likewise is the bias indicator enabled (prominently showing the political orientation of the current article).

6.5.6 Workflow and Questions

Our study consists of seven steps.[4] We refer to a *task set* as a sequence of steps associated with one topic, i.e., task set 1 refers to the first topic shown to a respondent, including the overview, the two article views, and respective questionnaires (steps 2–6). The *(1) pre-study questionnaire* asks for demographic and background data [332], such as age, political orientation, education, news consumption, and attitudes toward the topics we used [93]. *Generally, laws restricting abortion are [wrong ↔ right]. Generally, laws restricting the use of guns are [wrong ↔ right]. Generally, laws restricting environmental pollution are [wrong ↔ right].* For these questions and other questions concerning bi-polar adjective pairs, we use 10-point Likert scales. We also ask respondents whether the mentioned topics are relevant or irrelevant to them personally. Lastly, we ask whether they perceive the media to be biased against their views, in general, to better distinguish the treatment effects from prior skepticism (also called hostile media effect [282]).

Afterward, we show an *(2) overview* as described in Sects. 6.4.1 and 6.5.5 including instructions shown prior to the overview. The *(3) post-overview questionnaire* then operationalizes the bias-awareness in respondents (see Sect. 6.2.1) by asking about their perception of the diversity and disagreement in viewpoints, if the visualization encouraged them to contrast individual headlines, and how many perspectives of the public discourse were shown, e.g., *Do you think the coverage shown in the previous visualization represents all main viewpoints in the public discourse (independent of whether you agree with them or not) [not at all ↔ very much]? Overall, how did you perceive the articles shown in the previous visualization [very different ↔ very similar; very opposing ↔ very agreeing]?.* To match our definition of bias-awareness, we use as our main question (cf. [277]): *When viewing the topic visualization, did you have the desire to compare and contrast articles [not at all ↔ very much]?*

---

[4] To download the study materials, which contain all questions and exemplary screenshots of the visualizations, please refer to Sect. 6.1.

Afterward, we show an *(4) article view* as described in Sect. 6.4.2. A *(5) post-article questionnaire* operationalizes bias-awareness in respondents on an article level [332], i.e., *How did you perceive the presented news article? [very unfair ↔ very fair; very partial ↔ very impartial; very unacceptable ↔ very acceptable; very untrustworthy ↔ very trustworthy; very unpersuasive ↔ very persuasive; very biased ↔ very unbiased]*. We also ask whether the article contains political bias and biases against persons mentioned in the article. We repeat steps 2–5 two times since we show two task sets. After each overview, we showed two articles, i.e., we repeated steps 4 and 5 two times. To measure the effect of seeing an overview before an article, we also introduce a variant where we skip the overview. In such cases, the overview steps (2, 3) are skipped entirely. Afterward, a *(6) discrete choice question* asks respondents to choose between two articles, i.e., which one they consider to be more biased. In a *(7) post-study questionnaire*, respondents give feedback on the study, i.e., what they liked and disliked.

In the two pre-studies, where we tested the study design and usability of the visualizations (see Sect. 6.6), we repeated the same procedure with only one article after each overview and excluding step 5. In the first pre-study, we also excluded steps 2–6, since we only tested bias-sensitive overviews.

6.6 Pre-studies

Before our main study, we conducted two pre-studies (E1 and E2).[5] E1 consisted of 260 respondents recruited on MTurk (we discarded 3% from 268 respondents due to the quality criteria described in Sect. 6.5.4). E2 consisted of 98 respondents (we discarded 11% from 110).

The pre-studies aimed at testing the study design described in Sect. 6.5 and the usability of the visualizations described in Sect. 6.4. Further, we used the first pre-study to find a set of well-performing overviews, including representative baselines. This selection was necessary to satisfy the conjoint assumption "randomization of profiles." This assumption also requires that all profiles have the same set of attributes (Sect. 6.5.2). However, the number of attributes differs across our overviews (Sect. 6.5.5). For example, Plain has only two attributes (one for each headline tag), our bias-sensitive overview layout (Sect. 6.4.1) has an additional grouping attribute, and "no overview" naturally has no attributes. Thus, by determining which variants of the bias-sensitive layouts performed best in the pre-studies, we could fixate these overviews' attributes and compare them in the second pre-study and our main study.

Our first pre-study, E1, aimed to confirm the overall study design and collect effect data to make an informed selection of overview variants, both for the

---

[5] This section outlines the most important findings of our pre-studies. For more information concerning the pre-studies, please refer to [134].

PFA approach and the baselines. For the latter, we tested only variants using our bias-sensitive overview layout, where we randomly varied all attributes, i.e., grouping and the two headline tags. We identified (primarily insignificant) trends that indicated well-performing variants. In E2, we then tested the same design as planned for the main study (see Sect. 6.5.6), including article view and the other baselines (see below).

We also used the pre-studies to improve our design and visualizations. Reasons for partially mixed results in both pre-studies were various usability issues interfering with the effectiveness. For example, in E1, respondents reported they wanted to know how the grouping was performed and by whom. Before conducting E2, we addressed these shortcomings, e.g., by adding explanations (specific and generic) about how our system derives the classifications. After addressing these issues, we found positive, significant effects of our bias-sensitive overviews in the second pre-study, confirming our research design concerning the overview.

E2 revealed that showing both headline tags was most effective in improving bias-awareness in the Plain baseline. In contrast, for the bias-sensitive overviews, the bias-awareness remained unchanged or decreased if one or both tags were shown. We suspected that users might feel overwhelmed if many visual clues are present due to a higher cognitive load and potentially visual clutter. Further, the effect of the bias-sensitive layout itself was stronger than those of the headline tags if employed in a bias-sensitive layout. In sum, headline tags seemed to be most effective if employed in an otherwise bias-agnostic visualization, such as Plain.

Using the pre-study findings, we defined the following overview variants for the main study: *(1) No overview*; *(2) Plain* as described in Sect. 6.5.5; *(3) PolSides* as described in Sect. 6.5.5 with PolSides headline tags enabled to closely resemble the bias-sensitive news aggregator AllSides.com [8]; *(4) MFA* using the bias-sensitive layout (Sect. 6.4.1), Grouping-MFA (Sect. 6.3), and polarity headline tags enabled, which was the best-performing variant of MFA in our pre-studies; *(5) PolSides-generic* being identical to (3) but using generic explanations; *(6) MFA-generic* being identical to (4) but using generic explanations; *(7) Random-generic* using the bias-sensitive layout and random grouping; and *(8) ALL-generic* using the bias-sensitive layout, ALL-generic (Sect. 6.3), and cluster headline tags enabled. Note that we did not test a variant of Grouping-ALL with specific explanations.

In sum, we already found that the bias-sensitive overviews (PolSides and MFAP) yielded significant, positive effects on bias-awareness, confirming the overall study design. We also identified weaknesses, e.g., respondents criticized the lack of transparency regarding how the framing groups were determined and by whom. Before our main study, we addressed the identified shortcomings, e.g., by adding explanations about how our system derives the classifications. The study design and the visualization described in Sects. 6.4 and 6.5 are the results of our refinements and improvements using the pre-studies' findings. We also used the pre-studies to select a set of well-performing overview variants, including baselines, to be compared in the main study.

6.7 Evaluation

For our evaluation, we used the study design described in Sect. 6.5. In our main study, we recruited 174 respondents on MTurk, from which we discarded 8% using our quality measures. In sum, the $n = 160$ respondents (age: [23, 77], $m = 45.5$, gender (f/m/d): 72/88/0, all native speakers, political orientation (liberal (1)–conservative (10)): $m = 4.83, sd = 2.98$; see Appendix A.3) provided answers to 283 post-overview questionnaires (excluding "no overview"), 320 discrete choices on article views, and 640 post-article view questionnaires. The average study duration was 15 min ($sd = 6.22$). In the following, we present the results and discuss our findings for our primary study question regarding the overview and the secondary study questions concerning the article view and respondent factors (Sect. 6.5.1). If not noted otherwise, the reported effects were operationalized using the main question of the post-overview questionnaire and the additive score of all post-article questions (Sect. 6.5.6).

6.7.1 Overview

In our user study, the bias-sensitive overviews increased respondents' bias-awareness compared to the Plain baseline. PolSides achieved the highest effect when shown with specific explanations (Est. = 21.34). In the single-blind setting, i.e., if shown with generic explanations, the PolSides baseline had no significant effect (Est. = 8.46, $p = 0.17$).

In contrast, the PFA approach strongly and significantly increased bias-awareness in both settings: when specific explanations were used, our grouping method achieved a strong effect (MFA: Est. = 17.80). In the single-blind setting, **only the PFA approach consistently, significantly, and most strongly increased bias-awareness** (MFA, 13.35; ALL, 17.54).

Discussion of the Approaches and Their Results

But why is there a loss of effectiveness of PolSides in the single-blind setting, i.e., if generic explanations and labels are shown? Fully elucidating this question would require a larger sample size concerning respondents and topics. However, we qualitatively and quantitatively identified three potential, partially related causes, which we outline in the following.

(1) Popularity and Intuition The left-right dichotomy employed by PolSides is a well-known concept and easily understood by news consumers. None reported they did not understand the concept, and 20% of respondents exposed to PolSides praised that the bias concept, i.e., the grouping mechanism, was easy to understand, e.g., "I

Table 6.1 Shown are the effects on respondents' bias-awareness after overview exposure. Column "Est." shows the percentage increase in bias-awareness for the attributes CDCR, Overview, and Topic compared to their respective baselines, i.e., CoreNLP, Plain, and bushfire. Columns "SE," "z," and "p" refer to the standard error, z-score, and p-value, respectively. In column "p," asterisks represent the significance level where weakly significant ("*"), significant ("**"), and strongly significant ("***") refer to $p < 0.05$, $p < 0.01$, and $p < 0.001$, respectively

Attribute	Level	Est.	SE	z	p
CDCR	TCA	1.05	2.61	0.40	0.69
Overview	Random-generic	6.73	5.85	1.15	0.28
	PolSides	**21.34**	4.84	4.40	***
	MFA	**17.80**	4.90	3.63	***
	PolSides-generic	**8.46**	6.19	1.36	**0.17**
	MFA-generic	**13.35**	5.03	2.64	**
	ALL-generic	**17.54**	5.61	3.12	**
Topic	abortion law	0.78	2.96	0.26	0.79
	gun control	3.16	2.94	1.07	0.25

liked that it was laid out with left, center, right. It was intuitive." In contrast, PFA and its grouping techniques MFA and ALL are novel and somewhat technical, as are their (specific) explanations. For example, 40% of respondents exposed to MFA found its specific explanations (slightly) confusing and too "technical." In contrast, only 10% of respondents exposed to any of the generic variants, including MFA and ALL, reported comprehension issues. This improvement might lie in the generic explanations being less technical but more conceptual compared to the specific explanations of MFA.

These findings potentially indicate that a proportion of the bias-awareness effect in any visualization is due to encouraging users to look for frames and biases. Albeit not significant, the mild effects of the Random-generic overview (Est. = 6.73 as shown in Table 6.1) might serve as a rough approximation for the "base" effectiveness of bias-sensitive visualizations. This base effectiveness appears to be partially independent of the visualized content, such as the framing groups, and its meaningfulness. In practical terms, solely encouraging users to expect biases, e.g., in our study due to bias-sensitive layouts and explanations, can increase bias-awareness. Referring to intuitive or well-known bias concepts can improve this base effectiveness further, as indicated by the previously outlined effectiveness of PolSides that is only present with specific explanations. This effect of using well-known bias concepts is partially in line with our second finding (see afterward), i.e., the learning effect noticed for our novel bias and grouping concept.

(2) Learning Effect Our study indicates that the novel PFA approach might have benefitted in the course of the study from respondents' increasing understanding concerning how the approach works. Tables 6.2 and 6.3 show the bias-awareness effects after overview exposure in the study's first and second task sets. While we notice an effect increase for all overview variants in the second task set compared to

Table 6.2 Effects on bias-awareness after overview exposure in the first task set

Attribute	Level	Est.	SE	z	p
CDCR	TCA	-1.36	3.64	-0.37	0.71
Overview	Random-generic	6.88	7.49	1.23	0.22
	PolSides	**21.55**	6.36	3.39	***
	MFA	9.21	7.49	1.23	0.22
	PolSides-generic	6.69	7.28	0.92	0.36
	MFA-generic	6.97	6.60	1.06	0.29
	ALL-generic	9.39	6.63	1.42	0.16
Topic	abortion law	5.04	4.73	1.07	0.29
	gun control	5.77	4.56	1.27	0.21

Table 6.3 Effects on bias-awareness after overview exposure in the second task set

Attribute	Level	Est.	SE	z	p
CDCR	TCA	2.78	3.91	0.71	0.48
Overview	Random-generic	8.40	9.32	0.90	0.37
	PolSides	23.54	7.45	3.16	**
	MFA	**28.12**	7.03	4.00	***
	PolSides-generic	12.41	8.48	1.46	0.14
	MFA-generic	21.78	7.45	2.92	**
	ALL-generic	**26.46**	8.30	3.19	**
Topic	abortion law	-2.13	5.11	-0.42	0.68
	gun control	0.85	4.80	0.18	0.86

the first, there are key differences. First, while in task set 1, the PFA approach did not significantly increase bias-awareness, in the next task set 2, the PFA variants yielded the strongest, most significant effects. Specifically, PFA's and its **MFA grouping achieved the strongest effect among all overviews** (MFA, Est. = 28.12; PolSides, 23.54). Second, while in task set 1, only PolSides significantly increased bias-awareness, it could not benefit as much as the PFA approach from respondents' learning effects (effect increase from task set 1 to 2: MFA, =18.91pp.; PolSides, 1.99pp.).

These differences of the effects across the two task sets are also in line with the previously mentioned popularity and intuition (cause 1). Specifically, since the PFA approach is novel and its explanations are somewhat "technical," as respondents reported, we can expect both PFA's lack of effects in task set 1 and the learning effect throughout the study. Simultaneously, for the well-known and easy-to-understand left-right dichotomy employed by PolSides, we can expect significant effects from the beginning and only a slight increase of PolSides's effects throughout the study. These findings are also in line with the framing groups' substantiality discussed afterward (cause 3).

Table 6.4 Frames in the gun control event that we inductively identified when qualitatively analyzing the articles of individual framing groups determined by the grouping methods

Method	Frame	Headline
MFA	gun rights (emotional)	49 **killed** in shooting at Florida nightclub in possible act of Islamic terror
	gun control (factual)	**Support For Gun Control** Spikes After Orlando Shooting
	gun control (argumentative)	Orlando Shooting **Reignites Gun Control** Debate in Congress
PolSides or ALL	gun control (argumentative)	Orlando Shooting **Reignites Gun Control** Debate in Congress
	gun control (background)	Suspect **Purchased Guns Legally** Ahead Of Deadliest Shooting In Modern U.S. History
	gun rights (emotional)	**Father** Of Orlando **Victim: I Wish** Someone Else In That Club Had A Gun

(3) Substantiality of Framing Groups We qualitatively analyzed the groups yielded by individual grouping methods. We found that all methods, including PolSides, determined meaningful groups for most topics. For example, in the gun control topic, the groups yielded by any grouping method resemble the frames "gun control" and "gun rights" with subtle differences between the groups. Table 6.4 gives an overview of the frame we inductively identified when reading headlines and articles of each group. Note that while often frames were already apparent in the headlines shown in the table, in some cases, the groups' underlying frames more clearly emerged from reading the lead paragraph.[6] Grouping-MFA yielded two "gun control" frames (one was argumentative; the other used factual language) and one "gun rights" frame (focusing on cruelty and the shooter). Also, PolSides and Grouping-ALL yielded two "gun control" and one "gun rights" frames. Here, the "gun rights" frame determined by Grouping-ALL focused not on the shooter but on the victims and their right to defend themselves. In sum, all methods yielded framing groups of articles representing meaningful frames present in the coverage.

However, for some topics, the framing groups determined by MFA and ALL seemed to be more substantial compared to groups of PolSides. This finding is intuitive since PolSides determines the groups through the political orientation of the articles' outlets and thus is content-agnostic. In contrast, PFA analyzes an article's polarity toward individual persons, i.e., it uses in-text features for bias identification. Our respondent sample is too small for showing significant effects when sub-grouping for topics. However, the debt ceiling topic employed in our pre-studies highlights this methodological difference. As shown in Table 6.5, MFA yielded coherent groups that framed the deal positively by focusing on positive effects for the economy (frame 1) and countries' safety through the military (frame 2) or

---

[6] All articles, including their lead paragraphs and main text, can be found in the published study materials; see Sect. 6.1.

Table 6.5 Frames in the debt ceiling event that we inductively identified when qualitatively analyzing the articles of individual framing groups determined by the grouping methods

Method	Frame	Headline
MFA	positive (economical)	Trump, Congress Clinch **Debt-Limit** Deal After Tense Negotiations
	positive (political)	Donald Trump, congressional Democrats **reach two-year budget deal, avoid crisis** on debt ceiling
	negative (political)	Donald Trump and the G.O.P. Confirm **Their Fiscal Conservatism Was a Sham**
PolSides	positive (overview)	**Trump Announces Deal** On Debt Limit, Spending Caps
	positive (military)	Donald Trump, congressional Democrats **reach two-year budget deal, avoid crisis** on debt ceiling
	positive (military)	White House, congressional Democrats **agree on debt ceiling hike**

negatively, e.g., as political hypocrisy (frame 3). In contrast, the PolSides's groups—despite the topic has assumed left-right polarization—represent rather superficial frames. Specifically, all of PolSides's groups frame the deal positively, i.e., two groups frame the issue highly similarly, the other focuses on the overall implications of the deal. The issue of non-substantial frames is also present when analyzing coverage on events where the political left and right do not have distinct positions, as the "bushfire" event discussed below shows.

Of course, the reliability of our inductive frame analysis is lower than that of a deductive frame analysis performed by multiple annotators with high inter-annotator reliability. However, our qualitative findings concerning the substantiality are in line with the quantitative effects measured in our study. The effects in the single-blind setting show that PFA increased bias-awareness stronger than PolSides, both when looking at the overall study and the second task set. Since all approaches "look" identical in the single-blind setting, effect differences can only be explained by the framing groups the approaches determined. PolSides achieving the lowest increase in bias-awareness is thus intuitive. Moreover, Grouping-ALL achieved a larger increase than Grouping-MFA. This finding is also intuitive since we expect Grouping-ALL to be more reliable than Grouping-MFA, since the former uses more features (all persons) for determining framing groups than Grouping-MFA, which only uses the MFA's polarity.

None of the tested approaches yielded consistently meaningful framing groups and frames in coverage on the "bushfire" event. This finding is expected for the PFA approach, which identifies frames due to how persons are portrayed. However, in the "bushfire" event, much coverage focused on the fire's consequences for the economy or the environment. The finding is also intuitive for the PolSides approach since the political left and right do not have distinct positions in the "bushfire" event.

Criticism and Comments by Respondents

Lastly, we explored respondents' comments they provided in the post-study questionnaire to find issues and other trends in their comments. Table 6.6 shows a summary of the identified trends. Note that these qualitative findings are not representative[7] but can serve to get a preliminary understanding of advantages and issues respondents noticed.

Overall, 90% of the respondents who were exposed at least once to any bias-sensitive overview, including PolSides, explicitly mentioned finding the overview helpful for critically reviewing news coverage ("It also allows me to decide where I stand and learn new beliefs as I think that is important to see all sides of an article and to be able to understand it better" and "I like a trimmed down view that allows readers to efficiently compare articles."), compared to only 30% exposed to the Plain overview. For the MFA overview, 20% reported that they did not agree with the classification. However, 60% of respondents exposed to MFA reported that they liked quickly knowing the stance of an article, e.g., "I like that it gave you an idea on what stance the article had, whether it was pro, contra, or ambivalent." and "I liked that I could see different perspectives without having to go to different sources. I haven't seen anything much like that in any other app or online news sites. I think that this helps encourage critical thinking." The qualitative findings are further in line with the previously identified three causes for effect differences across the approaches. For example, the better substantiality of PFA's frames is reflected in the trend "Lack of substantiality or balance," which was 20% for PolSides and only 5% for Grouping-MFA and Grouping-ALL. Further, the technicality of PFA's explanations and the simplicity of left-right dichotomy are reflected in "Confusion about grouping mechanism," which was 0%, 40%, and 10% for PolSides, Grouping-MFA, and Grouping-ALL, respectively. Note that respondents reported less confusion for Grouping-ALL, which we tested only in the generic variant with less technical explanations, compared to the Grouping-MFA, which we tested also with specific explanations.

Summary

Overall, our study shows that bias-sensitive news overviews significantly and strongly improved bias-awareness in news consumers compared to popular, bias-agnostic news aggregators. Both the PFA approach and the PolSides baseline achieved positive effects. The PolSides approach even achieved the strongest effect when considering both task sets (Est. $= 21.34$), demonstrating the practical strength of this prior approach. However, PolSides lost its effectiveness entirely when employed in a single-blind setting. In contrast, the PFA approach achieved significant, consistent, and strong effectiveness, both when employed in the single-

---

[7] For example, in some cases, we could not map respondents' comments to specific visualizations since we showed them two overviews and, in total, four article views. When we could not unambiguously identify which visualization a respondent was referring to, we strictly discarded the comment.

Table 6.6 Non-representative, qualitative trends concerning advantages and issues of the overviews reported by respondents. The columns *PolSides* and *MFA* (Grouping-MFA) include specific and generic visualization variants. The column *ALL* refers the generic visualization variant employing the Grouping-ALL mechanism. All figures are in percent

Trend	PolSides	MFA	ALL
Frames are balanced	10	20	**30**
Lack of substantiality or balance	**20**	5	5
Confusion about grouping mechanism	0	**40**	10
Doubts about grouping mechanism	10	20	20
Grouping mechanism intuitive	**20**	0	0
Increased interest in coverage	30	35	20

blind setting (17.54) and else (17.80). Moreover, in the second half of the study, the PFA approach was by far the most effective means to increase bias-awareness in respondents (best PFA, Est. = 28.12; PolSides, 23.54; and in the single-blind setting: best PFA, 26.46; PolSides (insignificant), 12.41).

The respondents' comments and our effect comparison from the first to the second task set suggested that the bias-sensitive PolSides baseline initially benefited from its well-known and easy-to-understand bias identification method. While all approaches benefited from a learning effect during the study, the PolSides baselines—perhaps because already well-known—benefited only mildly. In contrast, the PFA approach was by far the most effective means to increase bias-awareness in the second half of the study.

Lastly, the results of our inductive frame analysis of the approaches' framing groups strengthened the previous findings. Albeit not representative, our qualitative analysis suggested that the groups determined by PFA more consistently represented meaningful and substantial frames than those yielded by PolSides. In the conclusion of this thesis, Sect. 7.1 practically demonstrates the findings of our evaluation on the example of news coverage that we introduced in Sect. 2.6 to practically demonstrate the research gap.

6.7.2 Article View

In the article view, only the in-text highlights significantly increased bias-awareness if between 5 and 9 of them were shown. Otherwise, the article view did not significantly increase respondents' bias-awareness. This section discusses potential causes for this lack of significant effects.

Table 6.7 shows the overall lack of bias-awareness effects measured using the rating questions of the post-article questionnaire. There was a weak but significant increase in bias-awareness of the highly polarizing "abortion law" event compared to the baseline event "bushfire" (Est. = 3.92).

Table 6.8 shows the article view's effects when operationalizing bias-awareness using the forced-choice question. Here, in-text highlights achieved stronger but still

Table 6.7 Shown are the effects on respondents' bias-awareness after article view exposure operationalized using the additive score of all rating questions in the post-article questionnaire. Column "Est." shows the percentage increase in bias-awareness for the attributes CDCR, in-text highlights, polarity context bar, MFAP headline tags (jointly with the MFAP article indicator as described in Sect. 6.5.5), PolSides headline tags (jointly with the PolSides article indicator), and the article's topic compared to their respective baselines, i.e., CoreNLP, disabled (four times), and bushfire. Columns "SE," "z," and "p" refer to the standard error, z-score, and p-value, respectively. In column "p," asterisks represent the significance level where weakly significant ("*"), significant ("**"), and strongly significant ("***") refer to $p < 0.05$, $p < 0.01$, and $p < 0.001$, respectively

Attribute	Level	Est.	SE	z	p
CDCR	TCA	-0.74	0.71	-1.04	0.29
In-text highlights	single-color	0.03	1.09	0.03	0.98
	2-colors	0.21	0.94	0.23	0.82
	3-colors	-0.89	0.93	-0.96	0.34
Polarity context bar	enabled	0.00	0.68	-0.41	0.99
Headline MFAP tags	enabled	-0.04	0.71	-0.06	0.95
Headline PolSides tags	enabled	-0.28	0.68	-0.41	0.68
Topic	abortion law	**3.92**	0.93	4.20	***
	gun control	1.34	0.81	1.66	0.96

Table 6.8 Effects on respondents' bias-awareness after article view exposure operationalized using the forced-choice question and distinguishing the color modes for in-text highlights

Attribute	Level	Est.	SE	z	p
CDCR	TCA	0.40	0.45	0.89	0.37
In-text highlights	single-color	2.04	5.85	0.35	0.73
	2-colors	5.08	5.41	0.94	0.35
	3-colors	4.43	5.60	0.79	0.43
Polarity context bar	enabled	0.33	4.30	0.08	0.94
Headline MFAP tags	enabled	0.49	4.13	0.12	0.91
Headline PolSides tags	enabled	1.10	4.57	0.24	0.81
Topic	abortion law	-0.27	0.60	-0.46	0.65
	gun control	0.03	0.55	0.06	0.95

insignificant effects, e.g., showing in-text highlights using two-color mode achieved (Est. = 5.08). Table 6.9 shows the same data but modeled with the count of in-text highlights instead of their color mode.[8] Likewise, this analysis yielded insignificant effects with the exception that if the article view showed between 5 and 9 in-text highlights, respondents' bias-awareness was significantly increased (Est. = 13.43).

---

[8] Because of our small respondent sample, we could not investigate both factors at the same time. Thus, we created two separate conjoint models, as shown in Tables 6.8 and 6.9.

Table 6.9 Effects on respondents' bias-awareness after article view exposure operationalized using the forced-choice question and distinguishing the count of in-text highlights

Attribute	Level	Est.	SE	z	p
CDCR	TCA	0.32	0.66	0.48	0.63
In-text highlights	1–4	7.43	7.02	1.06	0.29
	5–9	**13.43**	5.72	2.35	*
	10–14	7.39	5.75	1.28	0.20
	>14	10.46	6.75	1.55	0.12
Polarity context bar	enabled	0.91	4.32	0.21	0.83
Headline MFAP tags	enabled	0.31	4.15	0.07	0.94
Headline PolSides tags	enabled	0.95	4.54	0.21	0.83
Topic	abortion law	-1.03	1.69	-0.61	0.54
	gun control	-1.38	2.39	-0.58	0.56

We were also interested in whether showing an overview before a single article affected bias-awareness in the article view compared to showing none. However, likely due to the issues discussed in the following, including a too-small respondent sample size, we cannot elucidate this question. Specifically, showing an overview before the individual news articles had inconclusive effects. We found a significant, mild effect caused by only the MFA overview (Est. $= 4.64$, $p = 0.003$). Other overviews had no significant effects compared to not showing an overview.

Discussion of the Article View and Its Results

We discuss four factors, which are partially related to another, that may explain the overall lack of significant effects of the article view.

(1) Lack of multiple Perspectives in the Article View Bias is context-dependent and thus depends at least to some degree on relating and contrasting perspectives (Sect. 6.2.1). The underlying relativity of bias could be one reason why the overview (showing multiple articles and perspectives) achieved strong effects. In contrast, the article view (showing primarily a single article) achieved no significant effects overall, despite both employing the same techniques for bias identification. Most components of the article view communicate information about the given article. Only the headline tags and the polarity context bar allow to contrasting articles to some degree.

(2) User Experience (UX) Issues Besides the previously mentioned potential conceptual issue of article view, we identified various infrequent UX issues in respondents' feedback that we had not noticed in previous tests, including the pre-

Table 6.10 Non-representative, qualitative trends concerning article view issues as reported by respondents. All figures in percent

Trend	Overall Frequency
Further articles: too similar	8.33
In-text highlights: too many	13.33
In-text highlights: wrong / missing	16.67
In-text highlights: inaccurate	20.00
Article content	6.67
Useless MFA	6.67

studies. Table 6.10 shows a summary of the identified, non-representative trends.[9] For example, respondents reported that there were too many in-text highlights (13% of respondents, e.g., they felt "overwhelmed") or that, in their opinion, relevant mentions were missing (17%). This is in line with the lack of significant effects of these attributes (see, e.g., Table 6.9).

Overall, respondents did not criticize the polarity context bar, but we noticed two UX issues of the bar. In some cases, the bar placed the circles of multiple articles at the same position. In these cases, respondents could not see and compare the overlapping articles, rendering the functionality of the bar invalid. The issue mainly occurred when Grouping-MFA was used because MFA uses a single feature, i.e., the article's polarity toward the single MFA. Despite the MFA being the most frequently mentioned person among all articles, a few articles did not mention the MFA often or at all. Such articles had an increased likelihood of being placed at the same position in the polarity context bar. Further, some respondents might not have been aware of the hover functionality to view the articles' headlines. Those respondents only saw each article's polarity toward the event's MFA. As a consequence, they could not contrast the articles' headlines, which typically allow for getting a first understanding of articles' main slant and content.

(3) Inaccurate In-Text Highlights While also a UX issue, we discuss inaccurate in-text highlights separately since their root cause is PFA. The methods employed by PFA yield incorrect results in some cases despite their technically high classification performances (coreference resolution employed in target concept analysis, $F1_m = 88.7$; target-dependent sentiment classification employed in frame analysis, $F1_m = 83.1$). The consistent effects achieved by the overview visualizations demonstrated that the methods' classification performances are sufficiently high to classify biases at the article level reliably. However, at the same time, users were likely much more

---

[9] Note that these qualitative findings are not representative. For example, in some cases, we could not map respondents' comments to specific visualizations since we showed each respondent in total four article views during the study. When we could not unambiguously identify which visualization a respondent was referring to, we strictly discarded the comment from our analysis. The trends can, however, serve to get a first understanding of the issues respondents noticed.

Conservative Prime Minister Scott Morrison has pledged to "address issues around hazard reduction for national parks, dealing with land clearing laws, zoning laws and planning laws around people's properties and where they can be built."

Morrison has furthermore denied accusations that he has failed to acknowledge climate change during his tenure in office. But he also has remained steadfast in protecting the coal industry and the hundreds of thousands of families who rely on it to make ends meet.

Fig. 6.12 Screenshot of misclassified sentiment shown by in-text highlights (second sentence: "Morrison [...]")

sensitive to misclassified in-text highlights when viewing them individually in the article view (20% of respondents reported that the highlights were inaccurate; see Table 6.10). Figure 6.12 shows an example of in-text highlights with misclassified sentiment polarity.

(4) Study Design and Sample Size In typical news consumption, newsreaders actively choose which articles they want to read.[10] However, to adhere to the conjoint requirements, we had to present randomly chosen articles to them. This difference is not harmful per se, since the conjoint design also assumes that the effects of respondents' mixed interest in articles would cancel each other out given a large enough sample. The lack of significant effects shown in Table 6.7, however, might indicate that conjoint design is not optimal for evaluating the article view's effectiveness or that the sample size was too small given the diversity of articles and how respondents interacted with them.

Summary

In contrast to the news overviews, the article view only increased respondents' bias-awareness if between 5 and 9 in-text highlights were shown (Est. = 13.43). Figure 6.13 depicts an example of such in-text highlights using the two-color mode. Respondents' comments also indicated the principle usefulness of the visualization, e.g., "it [the article view and its in-text highlights] got me to think about the content of the article and the parts of the story it chose to focus on." Lastly, the results suggested that the forced-choice question can better operationalize bias-awareness for the article view showing only a single perspective than the rating questions.

Besides the in-text highlights, the other elements of the article view did not increase bias-awareness. In our view, the most likely reasons for the overall lack

---

[10] While this difference in principle also holds for the overview, we expect its influence to be weaker in the overview. Newsreaders are used to see events they might not be interested in a news overview. Further, since the overview shows smaller pieces of information from multiple articles instead of a large piece of information from a single article, there is an increased chance for catching users' interest in the overview. In the second pre-study, for example, some respondents had reported that they had wanted to read a different article or were not interested in those shown to them. Conversely, none of the respondents had reported they had wanted to see a different event in the overview.

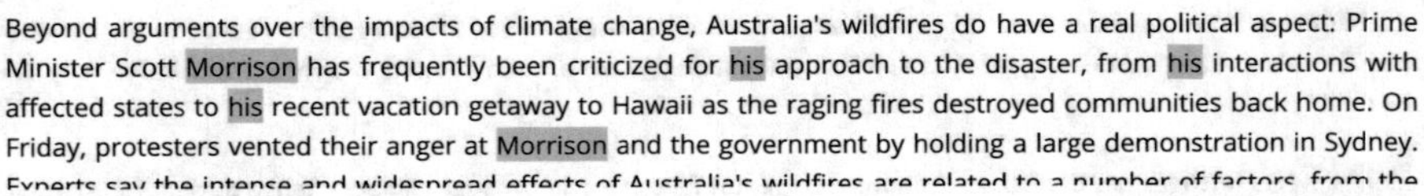

Beyond arguments over the impacts of climate change, Australia's wildfires do have a real political aspect: Prime Minister Scott Morrison has frequently been criticized for his approach to the disaster, from his interactions with affected states to his recent vacation getaway to Hawaii as the raging fires destroyed communities back home. On Friday, protesters vented their anger at Morrison and the government by holding a large demonstration in Sydney. Experts say the intense and widespread effects of Australia's wildfires are related to a number of factors, from the

Fig. 6.13 Screenshot of automatically classified sentiment in the article view's in-text highlights

of effects include the small sample of respondents and events, minor UX issues, and not showcasing multiple perspectives in the article view. In Sect. 6.8, we propose ideas to address the previously discussed issues.

6.7.3   Other Findings

Our study showed no significant effects of respondents' demographic and background factors on bias-awareness measured after overview or article view exposure. This finding contrasts prior studies, which indicate the influence of people's political orientation [20, 148], education, age, and other factors [183]. Like in the article view, one likely reason for the lack of significant effects in our study is the small respondent sample. Further, the distributions of most demographic and background attributes were imbalanced in our respondent sample (Sect. 6.7). While on the one hand, the distributions in our sample very roughly approximated those of the US population (cf. [183]), on the other hand, some attribute levels, such as in education, occurred too infrequently in our sample, even after dividing them into bins.

If, however, the resulting bins were sufficient in size, we found significant effects for some sub-groups, bins, and questions. For example, Table 6.11 shows the relative effects of a selection of respondents' demographic and other background attributes measured on the fourth post-overview question ("Overall, how do you think the coverage in the overview's articles compares to each other [very opposing ↔ very agreeing]"; see Sect. 6.5.6). Respondents having the lowest education level were significantly less aware of differences in the articles and perspectives compared to the baseline that had a bachelor's degree (Est. = −15.49). However, Table 6.11 shows no effects of other education levels and likewise were there only mixed effects in other post-overview and post-article view questions. Thus, due to the lack of effects similarly expected for other attributes and levels, the study's findings concerning respondents' demographic and background factors are only of exploratory nature.

Another finding was related to the hostile media effect (Sect. 2.2.3). The sub-group of respondents who generally perceived news as rather biased against their

Table 6.11 Effects of respondents' demographic and background attributes on post-overview question Q4

Attribute	Level	Est.	SE	z	p
Age	35–42	-4.87	3.54	-1.38	0.17
	43–52	-2.79	3.19	-0.88	0.38
	>53	0.64	4.22	0.15	0.88
Education	Some high school	**-15.49**	6.32	-2.45	*
	Some college credit	1.55	3.94	0.39	0.69
	Trade/tech./voc. training	0.88	4.74	0.19	0.85
	Master's degree	3.64	5.14	0.71	0.48
Political orientation	Center	**5.85**	2.97	1.97	*
	Conservative	0.43	3.55	0.12	0.90
Gender	Male	-0.44	2.90	-0.15	0.88

views[11] particularly benefited from seeing a bias-sensitive news overview with Grouping-MFA (Est. $= 2.86$, $p = 0.002$).

Our conjoint evaluation showed only a mildly positive, insignificant effect of using context-driven cross-document resolution compared to the CoreNLP-based baseline in the overview (Table 6.1) and mixed, insignificant effects in the article view (Tables 6.7 and 6.8). Potential reasons explaining the lack of an effect are as follows in our view. First, the overview relies on aggregated polarity information rather than individual mentions. Table 6.1 suggests that both the CoreNLP-based baseline and our method suffice to find and resolve mentions to substantially represent the articles' overall slant toward the most frequently mentioned persons. Second, while highlighting the polarity of individual mentions in the article view directly relies on high-quality individual mentions, the article view suffers from UX issues discussed in Sect. 6.7.2. Once these UX issues are addressed, we expect a larger sample size will yield a statistically significant effect of in-text highlights when using our coreference resolution due to its higher performance ($F1_m = 88.7$ compared to 81.9; see Sect. 4.3.6).

Summary

Our study showed no conclusive and significant effects of respondents' attributes, such as their political orientation and education level. This finding is in contrast with prior studies. We think that the main reason for this lack of expected tendency and influence is the respondent sample size. While sufficient to show consistent, strong, and significant effects across the news overviews, the sample size is too small for the relatively fine-grained attributes we queried in the background questionnaire. Moreover, the distributions of some of the respondents' attributes are imbalanced since they roughly approximate the distributions of the US population. In conjunc-

---

[11] The sub-group consisted of all respondents who answered the pre-study question "Most news is biased against my view [not at all (1) $\leftrightarrow$ very much (10)]" with greater equal 7.

tion with the small sample size, some attribute levels occur too infrequently to yield statistical soundness.

6.8   Future Work

We present ideas for future work to address limitations and issues concerning our approach and its evaluation. Specifically, we first discuss the limitations of our study design and the generalizability of the results. Afterward, we discuss the technical issues identified in our study.

Generalizability and Study Design

In our view, the main limitations of our experiments and results concern their representativeness and generalizability, mainly due to three partially related factors.

(1) Study Design For example, respondents had to view given events and articles rather than deciding what they read. An interactive design and a long-term observation study might more closely resemble real-world news consumption and address further issues that we not explicitly noticed but may have faced in our experiments, such as study fatigue. While our study's duration is well below where one would expect study fatigue [313], users on MTurk often work long and on many online tasks. Moreover, rather than querying respondents for the subjective concept of bias-awareness, a long-term observation study could directly measure the approaches' effects on news consumption, for example, whether respondents read more articles portraying events from different perspectives when using bias-sensitive visualizations [276].

(2) Respondent Sample While our sample roughly approximates the US distribution concerning dimensions important in this study, such as political orientation [282, 285], the sample contains selection biases, e.g., since we recruited respondents only on one platform and from only the USA. Thus, we cannot generalize our findings to news consumers from countries with systematically different political landscapes or media landscapes. For example, while in the USA, the two-party system has been shown to lead to more polarizing news coverage, countries with multi-party systems typically have more diversified media landscapes [395].

Further, we propose to increase the respondent sample to address the inconclusive or insignificant effects of article view, respondents' attributes, and when sub-grouping. Our sample size is larger than suggested by Cochran's formula [58, 385].[12] However, at the same time, the number of respondents was often too small in sub-groups or when analyzing respondents' demographic or other

---

[12] Strictly speaking, Cochran's formula suggests a minimum number of respondents, whereby our sample consists of fewer respondents, i.e., $n = 160 < 245$. However, they assume one observation per respondent, whereas we have two observations per respondent.

imbalanced attributes. Increasing the respondent sample would help yield overall better statistical soundness and allow for sub-grouping while retaining statistical significance.

Our *(3) event and article sample* yielded similar limitations as described previously for the respondent sample due to the event sample's small size and systematic creation. We thus propose increasing the number of events and articles per event. If both the event sample and respondent sample are increased, we could also use a random sample of events. A random event sample would reduce selection biases compared to a systematically selected sample. Lastly, our study did not relate bias-awareness to the articles' content, but only to our approach, respondents' attributes, and an event's expected degree of polarization as an approximation for the articles' amount of biases. To more precisely measure how articles' content and inherent biases influence bias-awareness, we propose to conduct a manual frame analysis and relate the identified frames to changes in bias-awareness.

The results of a manual frame analysis would also enable another way to evaluate bias identification approaches. Specifically, comparing the frames identified manually to those predicted automatically would allow for assessing the overall accuracy of an automated approach (Sect. 2.4). This research direction could even go as far as compiling benchmark collections of datasets from content analyses and frame analysis concerning media bias. Similar to the GLUE collection [373] and other benchmarks, such collections could be used to evaluate and directly compare the framing detection performance of individual approaches. Moreover, the benchmarks would also sharpen the approaches' representativeness and generalizability.

Technical Future Work Ideas

An idea from which effectiveness and UX of both the overview and the article view could benefit is to extract and show a distinct summary for each framing group. As a substitute for such a summary, the prototype currently shows the headline of a group's most representative article, allowing users to get an overall impression of that article's content and framing. However, a headline does not necessarily provide a concise summary of an article and even less so of the framing group that the article represents (Sect. 2.6). For example, one respondent reported that "[the] Overviews weren't long enough to get full scope of agreed-facts."

Besides the previously mentioned small respondent sample, other potential reasons for the lack of conclusive effects of the article view (except the in-text highlights) are rather technical. Addressing the view's UX issues is relatively straightforward. However, we think another reason for the lack of effects is the relativity of bias. On the one hand, our and other researchers' definitions of bias imply that bias requires contrasting information. On the other hand, the article view does not allow such comparison since it primarily shows a single article (with visual clues adding bias information). To address this conceptual issue, another line of research would need to be investigated. How can users be enabled to efficiently contrast individual "facts" presented in an article with matched facts from other articles (cf. [276])? We had excluded this idea when devising article visualizations for the study for various reasons. Most importantly, we think a visualization allowing for contrasting facts contradicts our ease-of-use objective (Sect. 6.4). For example, showing individual facts and alternative presentations of the same facts taken from other articles would increase the complexity of the article view. Further, this line of research requires the development of methods for news-specific semantic text similarity (Sect. 2.3.2). We have already taken the first steps toward this idea, and we proposed an exploratory system and visualization to explore how news articles use and reuse information from other sources [77, 139]. However, the preliminary evaluations of our exploratory approaches indicated the expected difficulty of this task.

To identify meaningful framing groups in non-person-centric coverage, such as on the "bushfire" event, we propose extending our analysis to further semantic concept types, such as groups of persons, countries, and objects. Our method for target concept analysis is already capable of resolving these types, and we would need only to extend the target-dependent sentiment classification method. Instead of analyzing framing effects, an automated approach could also investigate topic-independent frames or their derivatives (Sect. 5.1). Another idea to improve the distinctiveness of the resulting framing groups is to use a dynamic clustering technique in the frame analysis component, such as affinity propagation. We fixated this number to three using k-means to allow for direct comparison with the three groups by PolSides. However, by allowing for a variable number of groups, the groups determined by PFA could more closely match the characteristics of the data.

Outlook

The effectiveness of revealing biases in news articles raises crucial questions beyond the traditional scope of computer science. For example, one respondent of our study asked "How do you trust the ratings/categorization?" Of course, communicating how the individual "rating," i.e., perspective, was achieved is crucial, and we aimed to achieve this using the explanations added to all visualizations. However, the underlying issue is much more complex and vital. Who is to decide which sources and articles should be contained in the input set analyzed by automated approaches? Should extreme or alternative outlets be included in the analysis and visualization to increase the number of distinct perspectives? We think that answering these questions with further interdisciplinary research is a crucial prerequisite before using automated bias identification methods at scale in real-world news consumption.

6.9 Key Findings

This chapter presented the first system to identify person-oriented frames and reveal corresponding framing groups of articles in political event coverage. Earlier, such frames could reliably be identified only using high-effort methods, such as frame analysis as conducted in social science research or media literacy practices. We demonstrated the effectiveness of our person-oriented framing analysis (PFA) approach in a large-scale user study.

In the study's single-blind setting, we found that our approach and overview most strongly, significantly, and consistently increased respondents' bias-awareness. In particular, the PFA approach found biases that were indeed *present* in news articles reporting on person-centric events. In contrast, the tested prior work rather facilitated the visibility of potential biases, e.g., by distinguishing between left- and right-wing outlets. Other prior approaches suffer from analyzing single or shallow features (Sect. 6.2). Using such simple techniques can result in superficial or unmeaningful framing groups, as noted by multiple respondents, e.g., "I am not sure I agree (others, too) with many of the 'left/center/right' designations of the sources, as many were quite ambiguous and/or the topics at hand are not always agreed or disagreed upon concretely by members of the same political party." In sum, by using our methods for context-driven cross-document coreference resolution, target-dependent sentiment classification, and frame clustering on all persons mentioned in person-centric coverage, the PFA approach reliably identified frames and in turn achieved the strongest effectiveness.

Further, we discussed the limitations of our study, for example, regarding the findings' generalizability. Reasons include selection bias, e.g., the respondent sample consisted only of people located in the USA, and the event sample only of 30 articles in 3 news events. We propose to address these limitations by increasing and diversifying both samples. Another promising idea for future work is to adapt the study design to more closely resemble daily news consumption. Here, we propose to conduct a long-term study to observe how revealing biases affects news consumption directly.

The most substantial technical improvement idea would enable the PFA approach to identify meaningful framing groups also in non-person-centric coverage. To achieve this, we propose to extend PFA to analyze other concept types, such as groups of persons and countries. Since the other methods in PFA are already capable of analyzing these types, only the target-dependent sentiment classification method would need to be extended. Further, to address the overall lack of conclusive and significant effects of the article view, we propose investigating how the relative concept of media bias can more effectively be communicated by the article view while maintaining ease of use.

6.10 Summary of the Chapter

The work described in this chapter allows us to answer questions raised in earlier chapters. Further, we are now able to investigate the effectiveness of the individual components concerning the overall research question.

How does our method for coreference resolution employed in the target concept analysis component contribute to the overall success of our system?

Our conjoint evaluation showed only a mildly positive, insignificant effect of using context-driven cross-document resolution compared to the CoreNLP-based baseline (Tables 6.1 and 6.8). However, we expect to see increased effectiveness after addressing the identified UX issues in visualization components that directly rely on high-quality individual mentions, such as in-text highlights. Further, in the future, when extending PFA to analyze also concepts types other than individual persons, our coreference resolution can directly be used since it is already capable of resolving these types, partially with much higher performance than prior methods (Sect. 4.3.4.3).

How does our target-dependent sentiment method employed in the frame analysis component contribute to the overall success of our system? More specifically (as hypothesized in Sect. 3.3.2), does the focus of our research objective to consider only person-targeting forms of bias suffice to address the overall research question, which seeks to reveal substantial biases in daily news consumption effectively? Likewise, does analyzing the framing effects on the one-dimensional polarity scale instead of, for example, nuanced political frames suffice to tackle our overall objective (as questioned in Sects. 2.3.4 and 5.2.1)?

We found that focusing the analysis on persons and specifically on person-oriented polarity generally suffices to identify substantial perspectives present in coverage on policy issues effectively. In the study, our approach achieved overall high effectiveness. Once respondents got used to the concept of bias or our approach, i.e., in the second task set, the PFA approach led to the strongest increase in bias-awareness. Additionally, the qualitative investigation of all approach's resulting framing groups suggested that our approach found meaningful person-oriented frames in the analyzed articles. In contrast, prior approaches only facilitate the visibility of potential perspectives, as we also demonstrated for one approach in our study and in the practical demonstration of the research gap in Sect. 2.6.

However, PFA's effectiveness is limited by characteristics of the analyzed news coverage, especially whether news articles report on persons. Whereas the PFA approach increased bias-awareness if news coverage focused on the persons involved in an event, we found that it yielded groups of articles representing indistinct frames when applied to the "bushfire" topic. In this topic, news articles reported less on individual persons. Instead, the news articles primarily reported on the consequences for society, economy, and nature. The low performance for such non-person-oriented topics is expected due to the design of the approach, which relies fundamentally on mentions of persons (Sect. 3.3). Fully elucidating this question requires a larger and more diverse event sample.

One idea to detect meaningful frames and respective framing groups in non-person-oriented topics is to extend our target-dependent sentiment classification to classify the sentiment of additional target types, such as groups of persons, countries, and objects. Our method for cross-document coreference resolution is already capable of resolving such types (see Sect. 4.3.4). Another line of research, which resembles frame analyses used in the social sciences to systematically analyze media bias, is to classify frames or their more practice-driven derivatives, such as micro frames [193], our frame properties (see Sect. 5.2), or frame types

[45, 46]. However, each of them has its own limitations and challenges, typically requiring high annotation effort as we discussed in Sect. 3.3.2.

Chapter 7
Conclusion

Abstract This chapter concludes the doctoral thesis by summarizing the previously presented research (Sect. 7.1) and major contributions (Sect. 7.2). Lastly, the chapter discusses the limitations of the presented work and highlights ideas for future research (Sect. 7.3).

7.1 Summary

This dissertation proposed a novel, interdisciplinary approach to an open research problem in computer science, computational linguistics, and related disciplines, which is of pressing societal relevance: *revealing biases in news articles*, including subtle yet powerful bias forms such as source selection, including or omitting information, and word choice. Slanted news coverage, especially on policy issues, can have severely harmful effects, e.g., on public opinion and collective decision-making, such as in democratic elections [237, 259]. Revealing media bias to news consumers can help to mitigate these adverse bias effects and, for example, support news consumers in making more informed choices.

The first interdisciplinary literature review on media bias showed that automated approaches to reveal media bias so far suffer from providing only superficial or inconclusive results, albeit achieving high technical performance. Often, the approaches find technically significant but substantially irrelevant "biases" in news coverage. For example, an in-principle effective means to reveal biases is communicating the different slants present in news coverage to news consumers. However, prior approaches that aim to achieve this find perspectives that are technically different but do not represent meaningfully different perspectives. As a consequence, the approaches cannot effectively reveal biases. One key reason for their mixed results is that prior automated approaches analyze—or generally treat—bias as an only vaguely defined concept, for example, as

"subtle differences" [211],
"differences of coverage" [278],
"diverse opinions" [251], or
"topic diversity" [252].

© The Author(s) 2023
F. Hamborg, *Revealing Media Bias in News Articles*,
https://doi.org/10.1007/978-3-031-17693-7_7

The superficial methodology and non-optimal results become apparent when comparing the approaches to research in the social sciences. There, decade-long research on media bias has resulted in models to describe individual bias forms and effective methods to analyze them. The data-driven analyses determine, for example, substantial perspectives (also called *frames*) by identifying in-text means (also called *framing devices*) from which these frames empirically emerge. Due to their high effort and required expertise, these and other manual techniques do not reveal biases for the current day, which would be vital during daily news consumption to mitigate the severe bias effects.

To address the shortcomings of automated approaches and manual analyses, we devised *person-oriented framing analysis (PFA)*, a fundamentally different approach to bias identification. Compared to prior automated approaches, PFA does not treat media bias as a single, broadly defined concept but analyzes specific, person-oriented in-text means of bias to detect substantial perspectives indeed present in the analyzed news coverage. Compared to analyses conducted in social science research, PFA does not require manual reading and annotating news articles but automates these tasks entirely.

To automatically identify frames that could previously be reliably identified only using person-oriented, manual frame analyses, we designed PFA to imitate and automate this manual procedure. Specifically, PFA determines how individual news articles that are reporting on a given event portray the individual persons involved in the event. To achieve this, we introduced two components, which resemble tasks usually conducted by human annotators in person-oriented frame analyses. First, *target concept analysis* identifies in-text mentions of relevant subjects (here persons) that can be targeted by biases. The component then resolves the persons' mentions across all articles. Second, *frame analysis* determines how each mention's context (here sentence) portrays the respective subject. By subsequent clustering of articles that similarly portray the persons involved in the event, PFA is able to detect substantial frames and detect groups of articles having these frames.

To showcase the functionality and results of each component in PFA, we use a running example of news excerpts taken from three articles reporting on an event of the Republican Party's presidential primaries in 2016. We introduced these three and, in total, eight articles as part of our real-world example to practically demonstrate the research gap in our literature review (Sect. 2.6). Figure 7.1 shows the text excerpts.

We researched methods for the respective natural language understanding tasks of either component. For *target concept analysis*, we devised—among others— a dataset and method for context-driven cross-document coreference resolution in news articles. Coreference resolution aims to identify which mentions refer to which semantic concepts. In the presence of media bias, this is especially difficult and important since different journalists often refer to the same person, action, or other semantic concepts by using terms that are typically not synonymous or may even be contradictory in other contexts. Examples of such highly context-dependent coreferences include "intervene" and "invade" or "invading forces" and "coalition forces." Related techniques for coreference resolution capably resolve

After strong debate, Christie, Bush resume attack on Rubio	Here Are The Winners And Losers From ABC's Republican Presidential Debate	Marco Rubio is biggest loser. Trump and the governors all have a good night in NH
"He's a good guy, but he's just not ready to be president," Christie told "Fox News Sunday," after attacking Rubio for his inexperience in running government. "I felt justified because I've been saying this for a long time."	2) Donald Trump- [...] Trump made it clear that he doesn't want or need their money. Trump went back to selling winning, and if he wins on Tuesday, it is easy to see the Trump train getting on a roll.	With the exception of the moment when the crowd booed Trump [...] he was unwaveringly in charge. His answers on how important being an effective dealmaker [...] Trump was measured and thoughtful.
Trump was hit hard by Bush in the debate for his support of eminent domain, with Bush pointing out that Trump tried to take property for an "elderly woman" in Atlantic City for a casino parking lot.	2). Marco Rubio- This was a terrible debate for Rubio. The Senator from Florida came into the debate with momentum, and he ran into a buzzsaw named Chris Christie. Rubio keeps trying to sell himself as some [...]	It certainly helped that he wasn't that target on the stage since Marco Rubio filled that role on Saturday evening. [...] Rubio just moved into second place in New Hampshire – all eyes and attacks were on him. And he wasn't ready.

Fig. 7.1 Excerpts of articles from our real-world example introduced in Sect. 2.6. Each of the three boxes represents one article, and each of the two boxes within represents one paragraph. For simplicity, some paragraphs are skipped. Otherwise, the texts are unchanged, e.g., the order of the paragraphs is identical to the order in the articles, and typos are retained

After strong debate, Christie, Bush resume attack on Rubio	Here Are The Winners And Losers From ABC's Republican Presidential Debate	Marco Rubio is biggest loser. Trump and the governors all have a good night in NH
"He's a good guy, but he's just not ready to be president," Christie told "Fox News Sunday," after attacking Rubio for his inexperience in running government. "I felt justified because I've been saying this for a long time."	2) Donald Trump- [...] Trump made it clear that he doesn't want or need their money. Trump went back to selling winning, and if he wins on Tuesday, it is easy to see the Trump train getting on a roll.	With the exception of the moment when the crowd booed Trump [...] he was unwaveringly in charge. His answers on how important being an effective dealmaker [...] Trump was measured and thoughtful.
Trump was hit hard by Bush in the debate for his support of eminent domain, with Bush pointing out that Trump tried to take property for an "elderly woman" in Atlantic City for a casino parking lot.	2). Marco Rubio- This was a terrible debate for Rubio. The Senator from Florida came into the debate with momentum, and he ran into a buzzsaw named Chris Christie. Rubio keeps trying to sell himself as some [...]	It certainly helped that he wasn't that target on the stage since Marco Rubio filled that role on Saturday evening. [...] Rubio just moved into second place in New Hampshire – all eyes and attacks were on him. And he wasn't ready.

Fig. 7.2 Article excerpts after *target concept analysis*. Mentions of each candidate are underlined using candidate-specific style. For simplicity, mentions are not underlined of infrequent persons, such as "woman" in the left article, and other concept types our method can resolve, such as actions or objects

generally valid synonyms and nominal and pronominal coreferences, such as "Biden," "US president," and "he." However, current methods cannot reliably resolve the previously described mentions.

In contrast, our method extracts and resolves such mentions of persons and other semantic concepts across the input set of news articles. In the evaluation, our method achieved high performance for individual persons as analyzed by PFA ($F1_m = 88.7$) and thereby outperforms the state of the art ($F1_m = 81.9$). Figure 7.2 depicts the results of our method when applied to the running example. In addition to the mentions of the debate's candidates shown in the figure, our method found and resolved more context-dependent mentions, such as "the Republican candidate" (for Ted Cruz) and "the business man" (for Donald Trump).

For *frame analysis*, we devised—among others—the first method for target-dependent sentiment classification (TSC) in news articles. While researchers have developed TSC methods for various domains, state-of-the-art methods achieve "useless" [335] classification performance on the news domain ($F1_m \leq 44.0$). In the established TSC domains, such as product reviews or Twitter posts, authors express sentiment toward a target rather explicitly ("the camera is awesome"). In contrast, interpretation of sentiment in news articles requires a higher level of interpretation. For example, journalists typically express sentiment implicitly, such as by describing actions performed by a target person ("[...] Trump tried to take property for [from] an 'elderly woman' [...]," as shown in the left article of the running example). Prior methods for news TSC avoid this difficulty by focusing on cases with explicit connotations, such as direct quotes or readers' comments. However, by focusing on such infrequent cases, the methods neglect the majority content of an article and perform poorly when applied in real-world applications.

To enable TSC on news articles despite their implicit sentiment connotation and other differences compared to established TSC domains, we created *NewsMTSC*, the first dataset for news TSC consisting of over 11k labeled sentences. Afterward, we devised the first TSC model for the news domain. Our deep learning model achieves high TSC performance on news articles ($F1_m = 83.1$). The high performance allows for using our model in real-world news coverage as Fig. 7.3 shows for the running example. Our model resolved instances that prior news TSC approaches could resolve due to the sentences' explicitness ("He's a good guy [...]," left article; "Marco Rubio is biggest loser," right article). In contrast to prior approaches, our model also resolved the implicit connotations, such as "The Senator from Florida came into the debate with momentum [...]" (middle article) and "[...] Trump tried to take property for [from] an 'elderly woman' [...]" (left article).

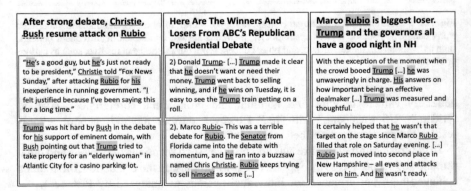

Fig. 7.3 Article excerpts after *target-dependent sentiment classification*. The candidates' mentions are colored depending on their local context's sentiment concerning the respective candidate: green or red for positive or negative sentiment, respectively. Mentions with neutral sentiment are not colored

Table 7.1 Results of approaches to identify biases in the real-world example. The columns "Headline," "Political," "Clustering," and "Frame" show each article's central perspective on the event according to the headline's potential frame, the outlet's political orientation, an automated clustering technique on word embeddings, and inductive manual frame analysis (ground truth). Please refer to Sect. 2.6 for more information on these approaches. Lastly, "PFA" shows the person-oriented framing groups as identified by the PFA approach. For each approach, the colors of its groups are chosen to maximize congruence with the framing groups of the ground truth. The higher the visual congruence of a column with the "Frame" column, the better

ID	Headline	Political	Clustering	Frame	PFA
0	H1	P1	C2	F2	PF1
1	?	P1	C2	F3	PF1
2	?	P2	C1	F1	PF0
3	H2	P2	C3	F2	PF2
4	H1	P2	C2	F3	PF1
5	H3	P3	C2	F2	PF2
6	H1(a)	P3	C2	F2	PF1
7	H1(b)	P3	C2	F3	PF0

Frame clustering completes the PFA approach. PFA employs a clustering technique, such as k-means, to group articles based on how each article portrays the event's persons. Table 7.1 shows the bias identification results for the eight articles of our real-world example introduced in Sect. 2.6, including the three used in this section's running example. The second to fourth columns represent prior approaches, both manual and automated. The fifth column ("Frame") represents the ground truth of framing groups derived using a standard tool in the social sciences, inductive frame analysis. Each cell shows the frame or "perspective" classified by the column's approach for the row's article.

None of the prior approaches yielded perspectives coherent to those of the inductive analysis. Albeit not perfect, only the PFA approach yielded person-oriented frames congruent to the frames derived from the inductive analysis. Of course, this running example cannot serve as an evaluation, nor is it intended to be. However, the results are representative of the weaknesses of prior work (see also Sect. 2.6) and the strengths of the PFA approach, as our evaluation (see also Sect. 6.9) summarized in the following shows. Before, we conclude our running example by introducing our prototype system.

As a proof of concept, we developed *Newsalyze*. Our prototype for bias identification and communication visualizes the results of the previously described methods. We devised non-expert visualizations designed to aid readers in daily news consumption. First, an *overview* aims to help users get a synopsis of news events and articles reporting on them. Subsequently, an *article view* shows a single news article, e.g., of an event of interest. We designed both visualizations to show not only the news content but also communicate biases present in the coverage. For example, established news aggregators show events and, for each event, one or more articles reporting on it. Our overview additionally reveals biases, for example, by showcasing articles of up to three person-oriented frames (part b in Fig. 7.4) and indicating an article's slant using a label (parts "a" and "b"). Such indicators are

Top tier takes heat: Rubio, others under fire at NH debate Perspective 3

The top tier in the Republican presidential race endured hard-hitting and sustained attacks on the debate stage Saturday night, with Florida Sen. Marco Rubio in particular getting pelted by New Jersey Gov. Chris Christie for skipping Senate votes and leaning on anti-Obama "talking points" on stage. [...]

> When reviewing media coverage on the given news topic, we identified three main perspectives. We determined the main perspective of each article. Below, you see for each perspective its most representative article.

Perspective 1	Perspective 2	Perspective 3
Rubio falters in presidential debate, offering hope to rivals	**Eighth GOP debate: Highlights from New Hampshire**	**GOP debate: Trump calls for a lot worse than waterboarding against terrorists**
Republican contender Marco Rubio struggled at a presidential debate on Saturday, potentially confounding his bid to emerge as Donald Trump's chief rival in New Hampshire. [...]	MANCHESTER, N.H. — Three days before the New Hampshire primary, the Republican candidates faced off on the debate stage at Saint Anselm College. [...]	An aggrieved Donald Trump was rejoining his Republican presidential rivals on the debate stage and hoping for a winning formula in New Hampshire. [...]

Further articles

▶ **Here Are The Winners And Losers From ABC's Republican Presidential Debate** Perspective 1

▶ **Marco Rubio is biggest loser. Trump and the governors all have a good night in NH** Perspective 2

▼ **Top tier takes heat: Rubio, others under fire at NH debate** Perspective 3

> The top tier in the Republican presidential race endured hard-hitting and sustained attacks on the debate stage Saturday night, with Florida Sen. Marco Rubio in particular getting pelted by New Jersey Gov. Chris Christie for skipping Senate votes and leaning on anti-Obama "talking points" on stage. [...]

Fig. 7.4 Screenshot of Newsalyze's news overview of news coverage on the US presidential primaries debate used in our real-world example. Similar to popular news aggregators, at the top, an article representative for the event coverage is shown (part "(**a**)") and at the bottom a list of further articles (part "(**c**)"). Additionally, Newsalyze shows a comparative part ("**b**") of up to three articles representative of the identified person-oriented frames. The comparative layout, the selection of these representative articles, and further bias-related information, such as the headline tags in the list of further articles, are intended to aid news consumers in quickly deciding which articles to read by understanding which articles offer which perspective

intended to help readers decide which articles to read, e.g., since they might offer different information on the event than already read articles.

Albeit the article excerpts shown in the overview do not directly summarize the frames, they typically allow for quickly grasping the essence of each frame. In our real-world example shown in Fig. 7.4, the headlines and lead paragraphs showcased as Perspectives 1–3 (in part "b") indicate the potential of Trump and other candidates (left article, e.g., "[...] offering hope to rivals. [...] confounding his bid to emerge as Donald Trump's chief rival [...]"]), the direct event coverage (middle, e.g., "Highlights from New Hampshire"), and the negative framing toward Trump (right, e.g., "Trump calls for a lot worse than waterboarding"). These three perspectives match to the frames of our inductive frame analysis conducted in Sect. 2.6.

To evaluate the effectiveness of revealing biases identified automatically by PFA in a real-world news consumption setting, we conducted a user study ($n = 160$ respondents). We implemented various baselines, e.g., one represented the bias-agnostic news overview of popular news aggregators, such as Google News. Another baseline resembled the bias-sensitive news aggregator AllSides, which relies on the left-right dichotomy. Specifically, AllSides determines the bias of an article by using its outlets' political orientation.

Showing the person-oriented frames identified by PFA increased bias-awareness in respondents significantly, strongly, and consistently. In contrast, the baselines increased bias-awareness not significantly or only under conditions. In a single-blind setting, i.e., where respondents did not know which bias identification method was used, only the PFA approaches significantly increased bias-awareness. Their relative effect on bias-awareness compared to the Google News baseline was Est. $= 17.5\%$ ($p < 0.002$). The AllSides baseline could significantly increase bias-awareness only if we revealed to respondents which bias identification method was used. If respondents were not informed, the same baseline achieved an insignificant effect, which was only slightly higher than the effect of a random baseline. In contrast, our PFA approach consistently led to strong and significant increases in bias-awareness. Moreover, respondents benefited from using our visualizations more than once. In the second half of the study, our PFA approach increased the bias-awareness most strongly (single-blind setting, Est. $= 26.5$ compared to the best baseline Est. $= 23.5$; if revealed, Est. $= 28.1$ compared to Est. $= 12.4$).

The study results and a qualitative investigation of the perspectives yielded by each approach indicated that the PFA approach effectively reveals biases by detecting meaningful perspectives indeed present in the news coverage. In contrast, the prior automated approaches suffered in some cases from their superficial methodology, which facilitates the detection of technically different but substantially irrelevant perspectives. To this extent, our study practically confirms the weaknesses of prior work as highlighted in our literature review.

Using technical means can only be part of a holistic solution to address media bias since it is ultimately news consumers who may be influenced by slanted news coverage. Thus, empowering news consumers to critically assess the news in order to mitigate adverse bias effects is, in our view, the higher-level goal. Effective means to achieve this include educating media literacy to news consumers and strengthening contrasting news formats, such as press review. However, such means typically require high effort. For example, even if news consumers have the skills to assess news coverage critically, researching alleged facts and contrasting event perspectives cause high effort. This tremendous effort may represent an insurmountable barrier to practically apply media literacy practices during daily news consumption. Automated approaches for bias communication can help to reduce the manual effort and thus represent a suitable means to enable critical assessment of coverage in daily news consumption.

Ultimately, the PFA approach can contribute to more informed decision-making. By enabling news consumers to effectively and effortlessly contrast substantial news perspectives, our approach contributes an effective means to support news consumers in critically assessing news coverage. We think that automated approaches for bias identification and communication are essential to enable bias-aware news consumption since only automated approaches can reduce the high manual effort required to contrast and critically assess news coverage.

7.2 Contributions

This section summarizes the contributions of this thesis for each of the research tasks presented in Sect. 1.3.

Research Task RT 1
Identify the strengths and weaknesses of manual and automated methods used to identify and communicate media bias and its forms.

To accomplish RT 1, we performed the first interdisciplinary literature review on media bias and approaches to analyze it as devised in the social sciences, computer science, and related disciplines. The review includes almost 200 research publications and related approaches. We found that automated bias identification approaches proposed so far often yield inconclusive or superficial results, especially compared to the results of decade-long research on the topic in the social sciences.

To facilitate interdisciplinary research on media bias, we established a shared conceptual understanding by mapping the state of the art from the social sciences to a framework, which approaches from computer science can target.

Research Task RT 2
Devise a bias identification approach that addresses the identified weaknesses of current bias identification approaches.

To overcome the deficiency of the current automated approaches for bias identification and communication, we introduced a novel approach named person-oriented framing analysis (PFA). Compared to prior automated approaches, PFA does not treat media bias as a single, broadly defined concept but analyzes in-text features representing specific person-oriented bias forms. To achieve this, PFA roughly resembles the manual process frame analysis as conducted by researchers in the social sciences. In contrast to these analyses, PFA does not require reading or annotating news articles but eliminates these tasks.

As a practical side contribution, we devised a system for crawling and extracting news articles from online outlets. The system can be used before PFA in order to gather news articles of interest. Additionally, the system has proven helpful throughout the research described in the thesis, e.g., to create training and test datasets.[1]

> **Research Task RT 3**
> *Develop methods for the devised approach and evaluate their technical performance.*

To detect person-oriented framing, we devised two analysis components. Our first component aims to identify persons and their mentions across a given set of news articles reporting on an event. Subsequently, our second component aims to determine how the individual persons are portrayed and then groups articles that portray the persons similarly, i.e., that frame the event similarly.

For the first component, we devised a method for context-driven cross-document coreference resolution that is the first to resolve also highly event-specific coreferences as they occur across slanted news articles, e.g., "Mr. Tough Guy" and "John Bolton" [230]. To evaluate this method, we created a test dataset of 50 news articles and 10 events. Each event contains articles by five outlets covering the political spectrum, including left-wing, center, and right-wing. When creating the dataset, we aimed to enable the annotation of also subtle and complex concept types, which the coreference resolution should resolve. To ensure reliable annotation despite the concept types' complexity, we conducted a manual content analysis as established in the social sciences. Our evaluation showed that our method reliably extracts and resolves individual persons ($F1_m = 88.7$ compared to $F1_m = 81.9$ achieved by the best baseline).

For the second component, we devised a method for target-dependent sentiment classification (TSC) in the domain of news articles. To evaluate this method, we created the first large-scale dataset for TSC in news articles reporting on policy issues. Our dataset consists of over 11k sentences. Each sentence includes at least one person mention and a sentiment label. Using an additional expert annotation on a random subset of the dataset, we confirmed the high quality of our ground truth. Not only increases our TSC method the state-of-the-art classification performance ($F1_m = 83.1$ compared to $F1_m = 81.7$), but our dataset is also the first to

---

[1] Our system is also used by numerous researchers to create pre-training datasets for deep learning language models [163, 387, 404], including the RoBERTa language model [214]. The tool has also shown helpful in COVID-19-related research [186, 301].

even enable TSC on the news domain under real-world conditions. News TSC was previously practically impossible due to the implicit character of sentiment in news articles compared to the established TSC domains.

Side contributions of our initial research to address RT 3 include a system that extracts 5W1H phrases describing the main event of an article, i.e., who did what, when, where, why, and how. Further, we devised and annotated so-called frame properties, which we had planned to use to categorize how sentences portray persons. Afterward, our exploratory research early showed a vital issue of automatically detecting substantial frames or derivatives: very high, initial annotation cost required to create a sufficiently large training dataset. Consequently, we devised the previously mentioned TSC approach to capture the fundamental effects of person-oriented framing, i.e., change of a person's sentiment polarity.

Research Task RT 4
Implement a prototype of a bias identification and communication system that employs the developed methods to reveal biases in real-world news coverage to non-expert news consumers.

To evaluate PFA in a setting that resembles real-world news consumption, we developed Newsalyze, a system for bias identification and communication. Besides integrating the previously devised bias identification methods into the system, we developed visualizations for non-expert news consumers. First, an overview allows for getting a synopsis of current events quickly. Second, an article view shows a single news article. We designed the visualizations to suit typical news consumption while complementarily showing bias-revealing information, such as the identified person-oriented frames present in the news coverage on a given event.

Research Task RT 5
Evaluate the approach's effectiveness in revealing biases by testing the implemented prototype in a user study.

To validate our approach's holistic and practical effectiveness in revealing biases, we conducted a large-scale user study ($n = 160$) on 30 news articles. Our studies measure the change in respondents' bias-awareness after exposure to one of our visualizations or baselines. We designed the studies to approximate real-world news consumption. Not only showed the study results that Newsalyze and the PFA approach significantly, consistently, and strongly increased bias-awareness in non-expert news consumers. By employing a conjoint design in our experiments, we were also able to pinpoint the effects of individual components in the visualizations. We used the conjoint design in the pre-studies, among others, to make a founded selection of strong visualization variants for the main study.

In addition to demonstrating the high effectiveness of PFA in a setting that resembles real-world news coverage, the evaluation led to the following conclusions:

- Not only does PFA increase bias-awareness. Our qualitative investigation of the resulting frames concluded that they are substantial: in contrast to the other approaches, frames detected by PFA in person-centric news coverage are consistently present in the coverage.
- Target-dependent sentiment classification is a fitting technique for the detection of person-oriented framing. This finding is contrary to the results presented in prior literature that suggested that the course, one-dimensional sentiment scale might not suffice for substantial bias analysis since it might fail to capture fine-grained nuances of framing.
- Since the PFA approach can detect substantial frames, we think it might be a suitable approach to complement the analyses conducted in social science research on media bias. There are differences between both use cases. For example, researchers in the social sciences pre-define frames for a specific research question and topic, whereas PFA implicitly defines the frames through its analysis and the resulting article groups. Nevertheless, the PFA approach might be readily usable for exploratory research. For example, researchers could inductively use PFA to detect frames in their data. In this scenario, PFA could serve as a replacement for inductive frame analysis, thereby reducing the manual effort in initial research phases.

7.3 Future Work

We intentionally formulated our research question rather openly and broadly to reflect the young state of the art in computer science. This way, our research question expresses the need to investigate how other disciplines define and analyze bias. Such disciplines traditionally include the social sciences, where media bias has been subject to research for decades resulting in comprehensive models to describe and effective methods to analyze it.

> How can an automated approach identify relevant frames in news articles reporting on a political event and then communicate the identified frames to non-expert news consumers to effectively reveal biases?

As part of the literature review, we established a shared conceptual understanding by mapping the state of the art from the social sciences to a framework, which computer science approaches can target. Using this framework and the identified weaknesses of prior bias identification and communication approaches, we nar-

rowed our open research question down to a specific research objective (Chap. 3).
Our objective focuses on the identification of person-oriented framing:

> Devise an approach to reveal substantial biases in English news articles
> reporting on a given political event by automatically identifying text-based,
> person-oriented frames and then communicating them to non-expert news
> consumers. Implement and evaluate the approach and its methods.

Concerning both the broad research question and the specific objective, this
section discusses the most important limitations of our research and derives future
research ideas to address these limitations.

7.3.1 Context-Driven Cross-Document Coreference Resolution

We devised the first method for context-driven cross-document coreference resolution. Naturally, our method can only serve as the first step in this novel task. Albeit
the method's design in principle allows for using the method outside the scope
of PFA, we focused on our research objective when devising and evaluating the
method. Our evaluation was not intended to and cannot elucidate how effective our
method is when applied to other domains and in other use cases. Before applying
our method in other use cases, we propose to conduct a more sophisticated evaluation, following all standards of coreference resolution evaluation. Specifically,
we propose to test the method on established datasets for coreference resolution
and compare it to a larger set of related methods. We also propose creating a
larger annotated dataset and thereby address various minor findings of the current
annotations.

A larger dataset would also enable the training of deep learning models for cross-
document coreference resolution. Recent models for single-document coreference
resolution achieve strongly increased performance compared to earlier traditional
methods [390]. We thus expect that such models could achieve higher performance
than our method employing hand-crafted rules. However, as described in Sect. 4.3,
the creation of a sufficiently large training dataset would cause high annotation cost.

7.3.2 Political Framing and Person-Independent Biases

Detecting political framing directly as opposed to detecting its effects on the framing
of persons (in PFA expressed through polarity) is beyond the scope of our objective
but relevant to the overall research question. While PFA represents an effective and
cost-efficient approach to identify substantial frames, its focus on persons implies
that the approach may fail if large parts of the coverage on an event are not person-

Table 7.2 Two news excerpts with opposing perspectives on the same event. The perspectives result from highlighting specific aspects of the event, here the negative consequences of loosening versus continuing COVID-19-related restrictions in Germany, March 2020. Free translation from [315, 354]

Publisher	Excerpt
tagesschau.de	**More than 37,000 infected persons in Germany**
	The number of registered coronavirus infected persons in Germany *has continued to rise*. The WHO *warns* against *lifting the corona measures* too early.
BILD.de	**Corona skepticism: We listen too much to virologists**
	A nation is sick when millions wrestle with a virus, but also when millions are consumed with *worry* that they will no longer be able to *feed their families*.

centric. Because of this expected shortcoming, we had initially attempted to identify topic-independent framing categories. Similar to other researchers' attempts to define or identify such categories directly, the approach turned out to be impractical. Reasons included the high cost of annotating a large dataset, which is required for training models to classify also nuanced and subtle framing categories.

Table 7.2 shows an example of framing that PFA cannot detect. Our approach would miss the two content-based frames, e.g., "restrictions are necessary" (first article) and "restrictions damage economy" (second article), because they are not directly related to individual persons. Instead, the articles focus on health or social implications versus economic consequences. To identify framing effects in non-person-oriented topics, we propose extending our approach to additional concept types, such as actions, person groups, countries, and objects. Our method for cross-document coreference resolution already resolves these types (Sect. 4.3.4). To classify the sentiment of these new types using target-dependent sentiment classification, we propose to extend the NewsMTSC dataset.

More closely resembling frame analyses as conducted in the social sciences, another idea is to determine frames directly rather than their effects. Key differences of person-oriented frames identified by PFA and political frames as defined by Entman [79] include that PFA defines person-oriented frames implicitly (each frame is defined by its articles) and inductively (frames are derived during the analysis). In contrast, social science researchers define frame for a specific research question before quantifying them in news coverage. In our view, this line of research represents the most refined approach, relying directly on decade-long, established social science research. Recent language models yielded a decisive performance leap in many natural language understanding tasks. However, training these models would cause high cost before they can classify subtle, analysis question-specific, and—viewed from a technical perspective—in part (intentionally) subjective frames (Sect. 6.10).

A pragmatic alternative to political frames might be topic-independent frame derivatives, such as microframes [193], our frame properties (see Sect. 5.2), or frame types [45, 46]. Each of these, however, has their limitations or challenges. Typically, they require manual validation or cause high annotation effort. Finally, active learning might be a suitable method to reduce the annotation cost. In active learning, human annotators need to label only a subset of examples. Typically, this subset consists of those examples the model to be trained is least sure how to classify. An iterative process of (re)training the model, selecting uncertain examples, and manually labeling them is repeated until the trained model achieves a sufficiently high classification performance.

Automated identification of framing effects or political frames directly could also be helpful in social science research. While frame analysis is an established means for analyzing how the media reports events and topics, the manual effort prevents conducting such analysis at scale. An automated approach for frame identification could assist researchers, especially in the early phases of their research. For example, PFA could serve as a tool to enable low-effort inductive exploration of news coverage by revealing which person-oriented frames are present in the data to be analyzed. Once extended to identify frames independent of persons or generally political frames, our approach could be helpful also in later phases of frame analyses.

Lastly, we propose to inspect the devised methods for biases they may have inherited from their training or fine-tuning data. We inspected both the annotated data and the model's predictions for structural biases indirectly, e.g., as part of the expert validation and manual error analysis. However, we did not directly probe the models and results for biases. This is advised, for example, because language models can be prone to gender-induced or other bias-related prediction errors due to their pre-training data [320]. Likewise, our fine-tuning data and the other datasets may contain structural biases despite the implemented quality measures. For example, the expert validation may entail biases since all experts were influenced by Western culture. Means to probe for biases in language models are already discussed in the literature (cf. [223]).

7.3.3 *Bias Identification and Communication*

This section uses the findings of our PFA evaluation to discuss both conceptual and technical means to address these limitations of the PFA approach and its evaluation.

In our view, an important limitation of the experiments and results is their generalizability, e.g., due to the study's design, its deployment, and the samples of respondents and events. For example, we measured respondents' bias-awareness using a set of questions. A future study could measure bias-awareness more directly, e.g., by observing respondents' news consumption behavior over a longer time frame. Like prior work, such a study would assume that effective bias communication encourages news consumers to view and compare more articles than bias-agnostic visualizations.

We also propose to sample respondents from other countries than the USA alone. Currently, we cannot generalize the study's findings to other countries' populations since political and media landscapes differ across countries. For example, while in the USA, the two-party system may lead to more polarizing news coverage, countries with multi-party systems typically have more diversified media landscapes [395]. We expect that these differences affect bias-awareness in general and thus also the effectiveness of our approach. Enabling studies in other countries requires extending the PFA approach to analyze news articles written in other languages. Ideas range from devising language-specific methods, where we expect overall high research effort and high-quality analysis results, to using neural machine translation before the PFA approach, which would require investigating the stability of task-specific linguistic properties, e.g., whether the subtle nuances of word choices are translated well.

The generalizability could also be improved by diversifying the event sample, which consisted in our study of 3 events and 30 articles. Besides systematically adding more events and articles, another future work idea is to randomly sample events and articles. While a random sample would reduce data selection biases, it would also require a larger respondent sample to compensate the noise introduced by the increased content diversity.

To further diversify the event sample, we propose investigating which technical changes are required to enable the PFA approach to analyze news from other sources than online newspapers. While we already included a few alternative outlets in our study, such as Breitbart, especially the news published on the increasingly consumed social media channels is different from news articles, such as concerning the texts' lengths and writing styles. We expect that the domains' differences require adapting the PFA approach. As a visionary outlook to broaden the generalizability of bias-awareness studies further, future approaches could aim to identify and reveal biases targeting not only persons. Ideas for extending the PFA approach range from analyzing other target concept types or political framing (as outlined previously in Sect. 7.3.2) to analyzing other forms of biases, such as through picture selection (Sect. 2.2.3).

We also propose to create a framing ground truth dataset using manual frame analysis. Such a dataset would enable measuring the influence of the articles' content and frames on respondents' bias-awareness. For example, by relating articles' content and frames with respondents' attitude toward these frames, the so-called hostile media effect could be investigated and how it affects the visualizations' effectiveness. Additionally, a framing ground truth would enable technical evaluation of the frame classification performance of approaches.

Our study showed effects of the PFA approach and the bias-sensitive overview but overall no effects of the article visualization and respondents' demographic factors. In the article view, only one visual clue conditionally increased bias-awareness.[2] Besides, the article view suffered from minor user experience (UX) issues and a conceptual issue, which lies in the relativity of bias. Specifically, to reveal bias, our article view should more prominently communicate frames of articles other than the viewed article. A method that could directly be integrated into our approach is to show summaries of other articles or their frames. Another future work idea is to enable users to contrast "facts" across articles, e.g., whether and how the facts of the currently viewed article are stated in other articles. To enable the mapping of such facts, methods for semantic textual similarity (STS) could be used. We already proposed exploratory systems for this line of research, but preliminary results indicate the difficulty of this approach [77, 139].

The respondent sample was too small to yield significant effects in the article visualization and respondents' attributes. On the one hand, the distributions of respondents' attributes, such as gender, political orientation, and education level, roughly approximated the US distribution. However, on the other hand, some distributions, such as of the education level, were thus imbalanced and contained too few observations in individual attribute levels, which prevented statistical analysis of these. Increasing the respondent sample would enable statistical analysis of the respondents' attributes. As discussed previously in this section, a larger respondent sample would also facilitate statistical soundness regarding the effect of the article view by compensating the article view's high content diversity.

7.3.4 Societal Implications

I conclude this thesis with an open outlook on the societal implications of using PFA and other automated approaches as part of daily news consumption. Understanding such societal implications is traditionally beyond the scope of computer science, and this dissertation cannot elucidate these implications. However, I intend to raise essential questions that could serve as a foundation for further, in-depth research across the topic's involved disciplines, especially in the social sciences.

- What are the causal relations of news consumption, readers' event assessment, and societal decisions? How can approaches for bias identification and communication sustainably support collective decision-making and other societal processes?
- Besides the previously mentioned model biases, what are other real-world pitfalls of using such automated approaches, and how can we prevent these pitfalls? For example, from all the outlets available, only a subset can be analyzed timely. Who could decide which sources automated approaches use for their bias analysis and

---

[2] The visual clue is highlighting persons' mentions in sentences of a given article. These in-text highlights use a different color depending on the sentences' sentiment toward the person; see Sect. 6.7.2.

subsequent visualization to news consumers? Which criteria are important for such selection?

The research described in this thesis highlights the effectiveness of taking an interdisciplinary approach to tackle the adverse issues caused by media bias. At the same time, the previously raised questions highlight the need for further interdisciplinary research on media bias. Not only is there a pressing need to do so, but now is also a perfect point in time. For example, news articles, representing the base of much research on media bias, are readily available in large quantities since much news is published online. Additionally, the rise of deep learning and recent language models has led to unprecedented advancements in natural language understanding. These advancements can and have enabled automated analysis of tasks that were previously difficult or nearly impossible to fulfill. This thesis, for example, devised the first model capable of classifying the subtle and implicit sentiment connotations common to news articles. I hope that with the shared conceptual understanding devised in the literature review and the proposed PFA approach, this thesis facilitates further interdisciplinary research on media bias.

My vision is that bias identification approaches, such as PFA, are integrated into popular news aggregators and news applications to enable bias-sensitive news consumption at scale. Enabling news consumers to critically assess and contrast news coverage can help mitigate the severe effects of systematically biased news coverage. Supporting individuals in making more informed choices is crucial in collective decision-making, such as democratic elections. In times of misinformation campaigns, "fake news," and other means to intentionally alter the public discourse, bias-sensitive news consumption is of unprecedented relevance. The PFA approach contributes to the required media literacy during daily news consumption by revealing substantial biases without tedious effort.

Appendix A

The Appendix includes further material related to the methods and prototype, an overview of publications receiving awards, and other resources.

A.1 Real-World Example

Table A.1.

Table A.1 Articles and their URLs used in the real-world example

ID	Outlet	Pol.	URL
0	CNN	L	https://edition.cnn.com/2016/02/06/politics/republican-debate-highlights/index.html
1	PoliticusUSA	L	https://www.politicususa.com/2016/02/06/here-are-the-winners-and-losers-from-abcs-republican-presidential-debate.html
2	USA Today	C	https://eu.usatoday.com/story/news/politics/onpolitics/2016/02/06/live-republicans-trump-cruz-rubio-new-hampshire-debate/79940814/
3	Chicago Tribune	C	https://www.chicagotribune.com/nation-world/ct-republican-debate-20160206-story.html
4	CNBC	C	https://www.cnbc.com/2016/02/06/trump-rubio-likely-targets-in-8th-republican-presidential-debate.html
5	Fox News	R	https://www.foxnews.com/politics/top-tier-takes-heat-rubio-others-under-fire-at-nh-debate
6	Fox News	R	https://www.foxnews.com/politics/after-strong-debate-christie-bush-resume-attack-on-rubio
7	Fox News	R	https://www.foxnews.com/opinion/marco-rubio-is-biggest-loser-trump-and-the-governors-all-have-a-good-night-in-nh

© The Author(s) 2023
F. Hamborg, *Revealing Media Bias in News Articles*,
https://doi.org/10.1007/978-3-031-17693-7

A.2　Context-Driven Cross-Document Coreference Resolution

This section contains additional information that complements the tables and numbers presented in the evaluation of our method for context-driven cross-document coreference resolution (Sect. 4.3.4.3) (Table A.2).

Table A.2 Overview of the normalized composition of the concept types across the events in NewsWCL50. Column "ID" refers to the events' IDs

ID	Action	Actor	Actor-I	Country	Event	Group	Misc	Object
0	8.5	37.4	2.3	35.5	12.5	0.0	3.4	0.0
1	0.0	44.1	7.5	20.4	6.5	0.0	3.7	17.6
2	7.4	24.3	1.3	37.9	9.7	0.0	16.7	2.4
3	0.0	9.2	2.0	14.8	14.1	35.2	24.5	0.0
4	0.0	35.6	10.8	25.6	3.1	0.0	8.2	16.4
5	4.2	29.7	10.3	22.3	27.9	0.0	0.0	5.4
6	2.2	4.1	16.1	25.0	0.0	32.9	6.8	12.0
7	0.0	6.7	6.5	51.4	0.0	1.9	11.4	21.8
8	0.0	31.9	26.6	15.1	11.0	4.5	0.0	10.6
9	0.0	16.6	7.8	35.5	3.0	0.0	17.0	19.8

A.3　Prototype Evaluation

Figures A.1, A.2, A.3, A.4, and A.5.

Fig. A.1 Gender distribution of the respondent sample in the main study

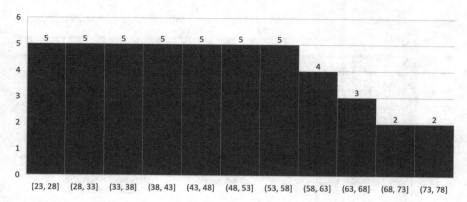

Fig. A.2 Age distribution of the respondent sample in the main study

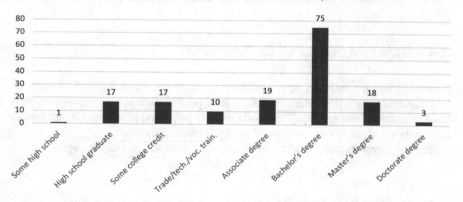

Fig. A.3 Education distribution of the respondent sample in the main study

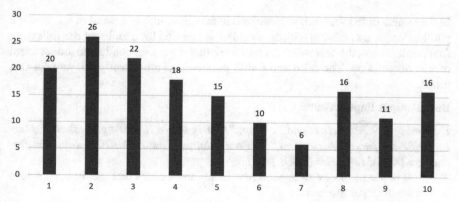

Fig. A.4 Political orientation distribution of the respondent sample in the main study, very liberal (1)–very conservative (10)

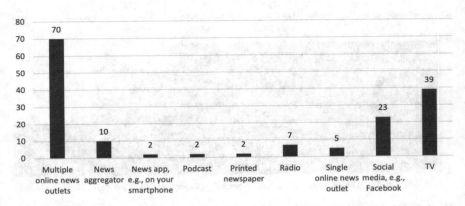

Fig. A.5 Distribution of the respondent sample in the main study on the attribute: news consumption

A.4   Data and Source Code Downloads

Following the principles of open access research, the author publishes all methods and materials after they were reviewed in a peer review process to ensure easy availability and high quality. Thus, the methods presented in this thesis and further resources are available on GitHub: https://github.com/fhamborg.

The prototype system Newsalyze is available at https://newsalyze.org/.

A.5   Publication Awards

In the course of the research summarized in this dissertation, several of the author's publications were honored with an award or nominated for one. From the following four publications, the first two represent work that the author conducted independent of this dissertation. The third and fourth publications are directly relevant for this thesis.

Best Student Paper Award

F. Hamborg, N. Meuschke, and B. Gipp, "Matrix-based News Aggregation: Exploring Different News Perspectives," in Proceedings of the ACM/IEEE Joint Conference on Digital Libraries (JCDL), 2017.
This full paper proposes an approach to tackle media bias that the author devised before his doctoral research.
http://jcdl.org/awards.php

Outstanding Paper Award

M. Schubotz, P. Scharpf, K. Dudhat, Y. Nagar, F. Hamborg, and B. Gipp, "Introducing MathQA: a Math-Aware question answering system," Information Discovery and Delivery, 2018.

Related to the overall field of information retrieval, this short paper introduces a question answering system for mathematical knowledge.

https://emeraldgrouppublishing.com/about/our-awards/emerald-literati-awards

Best Short Paper Award Nominee

F. Hamborg, A. Zhukova, and B. Gipp, "Illegal Aliens or Undocumented Immigrants? Towards the Automated Identification of Bias by Word Choice and Labeling," in Proceedings of the iConference 2019, 2019.

This short paper proposes an earlier variant of the person-oriented analysis approach that is summarized in Chap. 3.

https://ischools.org/Awards-2019

Best Full Paper Award Nominee

F. Hamborg, S. Lachnit, M. Schubotz, T. Hepp, and B. Gipp, "Giveme5W: Main Event Retrieval from News Articles by Extraction of the Five Journalistic W Questions," in Proceedings of the iConference 2018, 2018.

This full paper proposes an earlier variant of the event extraction system that is summarized in Sect. 4.2.

https://ischools.org/iConference-2018-Summary

Glossary

AMCE	Average marginal component effect. A term used in conjoint experiments to refer to the effect of a single attribute. See Sect. 6.5.2.
BERT	Bidirectional Encoder Representations from Transformers. A deep learning technique that achieved when published state-of-the-art or unprecedented performances on a variety of NLP tasks.
CAQDAS	Computer-assisted qualitative data analysis software Software tools that enable the annotation and optionally the analysis of texts and other documents.
CDCDCR	Context-driven cross-document coreference resolution. An admittedly rather long term coined by the author for a sub-type of coreference resolution where techniques resolve mentions across multiple documents while also linking mentions that may be considered contradictory in other contexts than the set of analyzed documents.
CDCR	Cross-document coreference resolution. A sub-type of coreference resolution where techniques resolve mentions across multiple documents.
CR	Coreference resolution. A set of techniques to resolve coreferential mentions. This process typically includes the identification of (relevant) mentions and linking those that refer to the same entity.
DC	Discrete choice. A term used in conjoint experiments to refer to the type of how respondents are asked their preference, i.e., here by being forced to decide for one variant from multiple.
EKS	External knowledge source. A term coined by the author to refer to dictionaries and other knowledge bases that are used as additional knowledge to gain a classification performance improvement in a language model. See Sect. 5.3.4.

© The Author(s) 2023
F. Hamborg, *Revealing Media Bias in News Articles*,
https://doi.org/10.1007/978-3-031-17693-7

F1	A performance metric that represents both quality and quantity of the results of an algorithm or other (automated) approach.
HAC	Hierarchical agglomerative clustering. Techniques to cluster, i.e., group, documents in a hierarchical way. See Sect. 2.3.1.
ICR	Inter-coder reliability. Level of agreement between multiple coders regarding their annotations done on the same set of documents. See also IRR.
IDF	Inverse document frequency. See TF-IDF.
IRR	Inter-rater reliability. In practice, often used synonymously to ICR. In this thesis, only used for labeling tasks, i.e., where given segments need to be assigned one or more codes. In contrast, in annotation tasks, also the segments have to be identified by the coders.
LM	Language model. A statistical machine learning model is a probability distribution over word sequences. In practice, these models are nowadays devised in the field of deep learning and are used to predict a desired output, e.g., sentiment polarity, given an input text.
MAgP	Mean average generalized precision. A performance metric used in multi-graded relevance assessments; see Sect. 3.5.2.
MFA	Most frequent actor. A term coined by the author to refer to the most frequently mentioned person or other semantic concept in a set of news articles reporting on a single event.
ML	Machine learning. A broad field of techniques that typically learn patterns from data and that can identify such patterns on new data. A sub-field of machine learning is deep learning, where artificial neural networks are employed.
MT	Machine translation. A set of techniques to automatically translate text from a source language to a target language.
MTurk	Amazon Mechanical Turk. A crowdworking platform.
NE	Named entity. A mention of a clearly defined entity, such as the name of a person or geographic location.
NER	Named entity recognition. Techniques to recognize named entities in texts.
NLP	Natural language processing. A broad field of techniques to process and analyze text documents.
NP	Noun phrase.
PD	Plagiarism detection. Techniques to detect the verbatim or disguised use of content or information from non-disclosed sources.
PFA	Person-oriented framing analysis. Approach used to identify frames in news articles. Term coined by the author, details explained in the thesis.
POS	Part of speech.
RT	Research task.

STS   Semantic text similarity. Techniques to measure the similarity of two text documents regarding their semantics, i.e., meaning.

TCA   Target concept analysis. Approach used in person-oriented framing analysis to first identify mentions of potential targets, such as persons, across all news articles. Term coined by the author, details explained in Chap. 4.

TF-IDF   Term frequency-inverse document frequency. A basic but well-established measure that reflects the importance of a word in a document given a set of documents.

TSC   Target-dependent sentiment classification. Techniques used to identify the sentiment, i.e., polarity, of a sentence or other context toward a target, such as a person.

VP   Verb phrase.

5W and 5W1H   The journalistic 5W questions and respective answers describe the main event of a news article, i.e., who did what, when, where, and why. 5W1H includes how the action was performed.

References

1. Sofiane Abbar et al. "Real-time recommendation of diverse related articles". In: *Proceedings of the 22nd international conference on World Wide Web*. ACM. 2013, pp. 1–12. DOI: 10.1145/2488388.2488390. URL: https://doi.org/10.1145/2488388.2488390.

2. Agence France-Presse. *Taliban attacks German consulate in northern Afghan city of Mazar-i-Sharif with truck bomb*. London, UK, 2016. URL: www.telegraph.co.uk/news/2016/11/10/taliban-attack-german-consulatein-northern-afghan-city-of-mazar/ (visited on 02/15/2021).

3. Charu C. Aggarwal and Jiawei Han. *Frequent Pattern Mining*. Ed. by Charu C.Aggarwal and Jiawei Han. Cham: Springer International Publishing, 2014. isbn: 978-3-319-07820-5. DOI: https://doi.org/10.1007/978-3-319-07821-2. arXiv: arXiv: 1011.1669v3. URL: http://link.springer.com/10.1007/978-3-319-07821-2.

4. Eneko Agirre et al. "SemEval-2015 Task 2: Semantic Textual Similarity, English, Spanish and Pilot on Interpretability". In: *Proceedings of the 9th International Workshop on Semantic Evaluation (SemEval 2015)*. Stroudsburg, PA, USA: Association for Computational Linguistics, 2015, pp. 252–263. DOI: https://doi.org/10.18653/v1/S15-2045. URL: http://aclweb.org/anthology/S15-2045.

5. Eneko Agirre et al. "SemEval-2016 Task 1: Semantic Textual Similarity, Monolingual and Cross-Lingual Evaluation". In: *Proceedings of the 10th International Workshop on Semantic Evaluation (SemEval-2016)*. Stroudsburg, PA, USA: Association for Computational Linguistics, 2016, pp. 497–511. ISBN: 978-1-941643-95-2. DOI: https://doi.org/10.18653/v1/S16-1081. URL: http://aclweb.org/anthology/S16-1081.

6. Phyllis F. Agran, Dawn N. Castillo, and Dianne G. Winn. "Limitations of data compiled from police reports on pediatric pedestrian and bicycle motor vehicle events". In: *Accident Analysis and Prevention* 22.4 (1990), pp. 361–370. ISSN: 00014575. DOI: https://doi.org/10.1016/0001-4575(90)90051-L.

7. Berfin Aktas, Tatjana Scheffler, and Manfred Stede. "Coreference in English OntoNotes: Properties and Genre Differences". In: Text, Speech, and Dialogue (TSD). Springer International Publishing, 2019, pp. 171–184. DOI: https://doi.org/10.1007/978-3-030-27947-9_15. URL: http://link.springer.com/10.1007/978-3-030-27947-9_15.

8. AllSides.com. AllSides - balanced news. 2021. URL: https://www.allsides.com/unbiased-balanced-news (visited on 02/24/2021).

9. Amanda Amos and Margaretha Haglund. "From social taboo to "torch of freedom": the marketing of cigarettes to women". In: *Tobacco control* 9.1 (2000), pp. 3–8. DOI: 10.1136/tc.9.1.3. URL: https://doi.org/10.1136/tc.9.1.3.

© The Author(s) 2023
F. Hamborg, *Revealing Media Bias in News Articles*,
https://doi.org/10.1007/978-3-031-17693-7

10. Jisun An et al. "Visualizing media bias through Twitter". In: Proc. ICWSM SocMedNews Workshop. 2012. URL: https://www.aaai.org/ocs/index.php/ICWSM/ICWSM12/paper/view/4775.

11. Nabiha Asghar. "Automatic Extraction of Causal Relations from Natural Language-Texts:AComprehensive Survey". In: arXiv preprint arXiv:1605.07895 (May 2016). arXiv: 1605.07895. URL: http://arxiv.org/abs/1605.07895.

12. FrederikAust et al. "Seriousness checks are useful to improve data validity in online research". In: *Behavior Research Methods* 45.2 (June 2013), pp. 527–535. ISSN: 1554–3528. DOI: https://doi.org/10.3758/s13428-012-0265-2. URL: http://link.springer.com/10.3758/s13428-012-0265-2.

13. Stefano Baccianella, Andrea Esuli, and Fabrizio Sebastiani. "SentiWordNet 3.0: An Enhanced Lexical Resource for Sentiment Analysis and Opinion Mining." In: *Proceedings of the Seventh International Conference on Language Resources and Evaluation (LREC'10)*. Vol. 10. Valletta, Malta: European Language Resources Association (ELRA), 2010, pp. 2200–2204. URL: https://www.aclweb.org/anthology/L10-1531/.

14. Brent H Baker, Tim Graham, and Steve Kaminsky. *How to identify, expose & correct liberal media bias*. Alexandria, VA: Media Research Center, 1994. isbn: 978-0962734823.

15. Eytan Bakshy, Solomon Messing, and Lada A Adamic. "Exposure to ideologically diverse news and opinion on Facebook". In: *Science* 348.6239 (2015), pp. 1130–1132. DOI: https://doi.org/10.1126/science.aaa1160. URL: https://science.sciencemag.org/content/348/6239/1130.

16. Alexandra Balahur et al. "Sentiment analysis in the news". In: *Proceedings of the Seventh International Conference on Language Resources and Evaluation (LREC'10)*. Valletta, Malta: European Language Resources Association (ELRA), 2010. URL: https://arxiv.org/abs/1309.6202.

17. Pablo Barberá et al. "Tweeting From Left to Right". In: *Psychological Science* 26.10 (Oct. 2015), pp. 1531–1542. ISSN: 0956-7976. DOI: https://doi.org/10.1177/0956797615594620. URL: http://journals.sagepub.com/doi/10.1177/0956797615594620.

18. Shany Barhom et al. "Revisiting Joint Modeling of Cross-document Entity and Event Coreference Resolution". In: Proceedings of the 57th Annual Meeting of the Association for Computational Linguistics. Stroudsburg, PA, USA: Association for Computational Linguistics, 2019, pp. 4179–4189. DOI: https://doi.org/10.18653/v1/P19-1409. URL: https://www.aclweb.org/anthology/P19-1409.

19. David P Baron. "Persistent media bias". In: Journal of Public Economics 90.1 (2006), pp. 1–36. DOI: https://doi.org/10.1016/j.jpubeco.2004.10.006.

20. Jonathan Baron and John T. Jost. "False Equivalence: Are Liberals and Conservatives in the United States Equally Biased?" In: *Perspectives on Psychological Science* 14.2 (Mar. 2019), pp. 292–303. ISSN: 1745–6916. DOI: https://doi.org/10.1177/1745691618788876. URL: http://journals.sagepub.com/doi/10.1177/1745691618788876.

21. Matthew A. Baum and Tim Groeling. "New Media and the Polarization of American Political Discourse". In: *Political Communication* 25.4 (2008), pp. 345–365. DOI: https://doi.org/10.1080/10584600802426965.

22. Eric P. S. Baumer et al. "A Simple Intervention to Reduce Framing Effects in Perceptions of Global Climate Change". In: *Environmental Communication* 11.3 (May 2017), pp. 289–310. ISSN: 1752–4032. DOI: https://doi.org/10.1080/17524032.2015.1084015. URL: https://www.tandfonline.com/doi/full/10.1080/17524032.2015.1084015.

23. Cosmin Bejan and Sanda Harabagiu. "Unsupervised Event Coreference Resolution with Rich Linguistic Features". In: *Proceedings of the 48th Annual Meeting of the Association for Computational Linguistics*. Uppsala, Sweden: Association for Computational Linguistics, 2010, pp. 1412–1422. URL: https://www.aclweb.org/anthology/P10-1143/.

24. Dan Bernhardt, Stefan Krasa, and Mattias Polborn. "Political polarization and the electoral effects of media bias". In: *Journal of Public Economics* 92.5-6 (June 2008), pp. 1092–1104. ISSN: 00472727. DOI: https://doi.org/10.1016/j.jpubeco.2008.01.006. URL: https://linkinghub.elsevier.com/retrieve/pii/S0047272708000236.

25. Timothy Besley and Andrea Prat. "Handcuffs for the Grabbing Hand? Media Capture and Government Accountability". In: *American Economic Review* 96.3 (May 2006), pp. 720–736. ISSN: 0002-8282. DOI: https://doi.org/10.1257/aer.96.3.720. URL: https://pubs.aeaweb.org/doi/10.1257/aer.96.3.720.

26. Clive Best et al. *Europe Media Monitor - System Description.* Tech. rep. December. 2005, pp. 1–57. URL: https://publications.europa.eu/flexpaper/common/view.jsp?doc=c0d6bb93-7ec4-496f-b857-b7fe9bc33d19.en.PDF.pdf&format=pdf&page=10.

27. Steven Bethard and James H Martin. "Learning semantic links from a corpus of parallel temporal and causal relations". In: *Proceedings of the 46th Annual Meeting of the Association for Computational Linguistics on Human Language Technologies: Short Papers.* Columbus, Ohio, USA: Association for Computational Linguistics, 2008, pp. 177–180. URL: https://www.aclweb.org/anthology/P08-2045.

28. Plaban Kumar Bhowmick. "Reader Perspective Emotion Analysis in Text through Ensemble based Multi-Label Classification Framework". In: Computer and Information Science 2.4 (Oct. 2009), pp. 64–74. ISSN: 1913-8997. DOI: 10.5539/cis.v2n4p64. URL: http://www.ccsenet.org/journal/index.php/cis/article/view/3872.

29. Douglas Biber and Edward Finegan. "Styles of stance in English: Lexical and grammatical marking of evidentiality and affect". In: *Text - Interdisciplinary Journal for the Study of Discourse* 9.1 (1989). ISSN: 0165-4888. DOI: https://doi.org/10.1515/text.1.1989.9.1.93. URL: https://www.degruyter.com/doi/10.1515/text.1.1989.9.1.93.

30. DavidMBlei. "Probabilistic topic models". In: Communications of the ACM 55.4 (2012), pp. 77–84. DOI: https://doi.org/10.1145/2133806.2133826.

31. Pablo J. Boczkowski. "The Processes of Adopting Multimedia and Interactivity in Three Online Newsrooms". In: *Journal of Communication* 54.2 (June 2004), pp. 197–213. ISSN: 0021-9916. DOI: https://doi.org/10.1093/joc/54.2.197. URL: http://joc.oupjournals.org/cgi/doi/10.1093/joc/54.2.197.

32. Tolga Bolukbasi et al. "Man is to computer programmer as woman is to homemaker? Debiasing word embeddings". In: *Advances in Neural Information Processing Systems.* 2016, pp. 4349–4357. URL: https://arxiv.org/abs/1607.06520.

33. Jelle Boumans et al. "The Agency Makes the (Online) News World go Round: The Impact of News Agency Content on Print and Online News". In: *International Journal of Communication* 12 (2018), pp. 1768–1789. URL: https://.oc.org/index.php/.oc/article/view/7109.

34. Dylan Bourgeois, Jérémie Rappaz, and Karl Aberer. "Selection Bias in News Coverage: Learning it, Fighting it". In: *Companion of the The Web Conference 2018 on The Web Conference 2018 - WWW '18.* 2018. ISBN: 9781450356404. DOI: https://doi.org/10.1145/3184558.3188724.

35. Tom Brown et al. "Language Models are Few-Shot Learners". In: 33 (2020). Ed. by H. Larochelle et al., pp. 1877–1901. URL: https://proceedings.neurips.cc/paper/2020/file/1457c0d6bfcb4967418bfb8ac142f64a-Paper.pdf.

36. TomÁš Brychcín and Lukáš Svoboda. "UWB at SemEval-2016 Task 1: Semantic Textual Similarity using Lexical, Syntactic, and Semantic Information". In: *Proceedings of the 10th InternationalWorkshop on Semantic Evaluation (SemEval-2016).* 2016. ISBN: 9781941643952. DOI: https://doi.org/10.18653/v1/S16-1089.

37. Hans Jürgen Bucher and Peter Schumacher. "The relevance of attention for selecting news content. An eye-tracking study on attention patterns in the reception of print and online media". In: *Communications* 347–368.31 (2006), p. 3. ISSN: 03412059. DOI: https://doi.org/10.1515/COMMUN.2006.022.

38. Ceren Budak, Sharad Goel, and Justin M. Rao. "Fair and balanced? Quantifying media bias through crowdsourced content analysis". In: *Public Opinion Quarterly* 80.S1 (2016), pp. 250–271. DOI: https://doi.org/10.1093/poq/nfw007.

39. C Bui. "How online gatekeepers guard our view: News portals' inclusion and ranking of media and events". In: *Global Media Journal* 9.16 (2010), pp. 1–41. URL: https://www.globalmediajournal.com/peer-reviewed/howonline-gatekeepers-guard-our-viewnews-portals-inclusion-and-rankingof-media-and-events-35232.html.

40. Business Insider. *These 6 Corporations Control 90% Of The Media In America.* 2014. URL: http://www.businessinsider.com/these-6-corporationscontrol-90-of-the-media-in-america-2012-6 (visited on 01/13/2021).

41. Carlos Busso et al. "Analysis of emotion recognition using facial expressions, speech and multimodal information". In: *Proceedings of the 6th international conference on Multimodal interfaces - ICMI '04.* New York, New York, USA: ACM Press, 2004, p. 205. ISBN: 1581139950. DOI: https://doi.org/10.1145/1027933.1027968. URL: http://doi.acm.org/10.1145/1027933.1027968.

42. Aylin Caliskan, Joanna J. Bryson, and Arvind Narayanan. "Semantics derived automatically from language corpora contain human-like biases". In: *Science* (2017). ISSN: 10959203. DOI: https://doi.org/10.1126/science.aal4230. arXiv: 1608.07187.

43. Erik Cambria et al. "SenticNet 4: A Semantic Resource for Sentiment Analysis Based on Conceptual Primitives". In: *Proceedings of COLING 2016, the 26th International Conference on Computational Linguistics: Technical Papers.* Osaka, Japan, 2016, pp. 2666–2677. URL: https://www.aclweb.org/anthology/C16-1251/.

44. Joseph N . Cappella and Kathleen Hall Jamieson. *Spiral of cynicism: The press and the public good.* Oxford University Press on Demand, 1997.

45. Dallas Card et al. "Analyzing Framing through the Casts of Characters in the News". In: *Proceedings of the 2016 Conference on Empirical Methods in Natural Language Processing.* Stroudsburg, PA, USA: Association for Computational Linguistics, 2016, pp. 1410–1420. DOI: https://doi.org/10.18653/v1/D16-1148. URL: http://aclweb.org/anthology/D16-1148.

46. Dallas Card et al. "The Media Frames Corpus: Annotations of FramesAcross Issues". In: *Proceedings of the 53rd Annual Meeting of the Association for Computational Linguistics and the 7th International Joint Conference on Natural Language Processing (Volume 2: Short Papers).* Stroudsburg, PA, USA: Association for Computational Linguistics, 2015, pp. 438–444. DOI: https://doi.org/10.3115/v1/P15-2072. URL: http://aclweb.org/anthology/P15-2072.

47. Xavier Carreras and Llu´is Màrquez. "Introduction to the CoNLL-2005 shared task: Semantic role labeling". In: *Proceedings of the Ninth Conference on Computational Natural Language Learning.* Association for Computational Linguistics. 2005, pp. 152–164. URL: https://www.aclweb.org/anthology/W05-0620/.

48. Angel X Chang and Christopher D Manning. "SUTime". In: *Climate Change 2013 - The Physical Science Basis.* Ed. by Intergovernmental Panel on Climate Change. iii. Cambridge: Cambridge University Press, 2012, pp. 1–30. DOI: https://doi.org/10.1017/CBO9781107415324.004. URL: http://www-nlp.stanford.edu/pubs/lrec2012-sutime.pdf.

49. Wei-Fan Chen et al. "Learning to Flip the Bias of News Headlines". In: *Proceedings of the 11th International Conference on Natural Language Generation.* Stroudsburg, PA, USA: Association for Computational Linguistics, 2018, pp. 79–88. DOI: https://doi.org/10.18653/v1/W18-6509. URL: http://aclweb.org/anthology/W18-6509.

50. Xiao Chen et al. "Aspect Sentiment Classification with Document-level Sentiment Preference Modeling". In: *Proceedings of the 58th Annual Meeting of the Association for Computational Linguistics.* Stroudsburg, PA, USA: Association for Computational Linguistics, 2020, pp. 3667–3677. DOI: https://doi.org/10.18653/v1/2020.acl-main.338. URL: https://www.aclweb.org/anthology/2020.aclmain.338.

51. Kyunghyun Cho et al. "Learning Phrase Representations using RNN Encoder- Decoder for Statistical Machine Translation". In: *Proceedings of the 2014 Conference on Empirical Methods in Natural Language Processing (EMNLP).* Stroudsburg, PA, USA: Association for Computational Linguistics, 2014, pp. 1724–1734. DOI: https://doi.org/10.3115/v1/D14-1179. URL: http://aclweb.org/anthology/D14-1179.

52. Darrell Christian et al. *The Associated Press Stylebook and Briefing on Media Law.* Basic Books, 2019. ISBN: 978-1541699892.

53. Darrell Christian et al. *The Associated Press stylebook and briefing on media law*. The Associated Press, 2014.

54. Junyoung Chung et al. "Empirical Evaluation of Gated Recurrent Neural Networks on Sequence Modeling". In: (Dec. 2014). arXiv: 1412.3555. URL: http://arxiv.org/abs/1412.3555.

55. Domenic V. Cicchetti and Alvan R. Feinstein. "High agreement but low kappa: II. Resolving the paradoxes". In: *Journal of Clinical Epidemiology* 43.6 (Jan. 1990), pp. 551–558. ISSN: 08954356. DOI: https://doi.org/10.1016/0895-4356(90)90159-M. URL: https://linkinghub.elsevier.com/retrieve/pii/089543569090159M.

56. Kevin Clark and Christopher D. Manning. Deep Reinforcement Learning for Mention-Ranking Coreference Models. Austin, US, 2016. DOI: https://doi.org/10.18653/v1/D16-1245. URL: https://www.aclweb.org/anthology/D16-1245.

57. Kevin Clark and Christopher D. Manning. "Improving Coreference Resolution by Learning Entity-Level Distributed Representations". In: *Proceedings of the 54th Annual Meeting of the Association for Computational Linguistics (Volume 1: Long Papers)*. Stroudsburg, PA, USA: Association for Computational Linguistics, 2016, pp. 643–653. DOI: https://doi.org/10.18653/v1/P16-1061. URL: http://aclweb.org/anthology/P16-1061.

58. William G. Cochran. "The Combination of Estimates from Different Experiments". In: *Biometrics* 10.1 (Mar. 1954), pp. 101–129. ISSN: 0006341X. DOI: https://doi.org/10.2307/3001666. URL: https://www.jstor.org/stable/3001666?origin=crossref.

59. Nicole S. Cohen. "AtWork in the DigitalNewsroom". In: *Digital Journalism* 7.5 (May 2019), pp. 571–591. ISSN: 2167-0811. DOI: https://doi.org/10.1080/21670811.2017.1419821. URL: https://www.tandfonline.com/doi/full/10.1080/21670811.2017.1419821.

60. Steven R Corman et al. "Studying Complex Discursive Systems." In: *Human communication research* 28.2 (2002), pp. 157–206.

61. Jackie Crossman. *Aussies Turn To Social Media For News Despite Not Trusting It As Much*. Nov. 2014. URL: https://www.bandt.com.au/aussies-turnsocial-media-news-despite-trusting-much/ (visited on 12/11/2020).

62. Agata Cybulska and Piek Vossen. "Using a sledgehammer to crack a nut? Lexical diversity and event coreference resolution". In: Proceedings of the Ninth International Conference on Language Resources and Evaluation (LREC'14). Reykjavik, Iceland: European Language Resources Association (ELRA), 2014, pp. 4545–4552. URL: https://www.aclweb.org/anthology/L14-1646/.

63. Christian S. Czymara and Marijn van Klingeren. "New perspective? Comparing frame occurrence in online and traditional news media reporting on Europe's "Migration Crisis"". In: *Communications* (Apr. 2021), pp. 1–27. ISSN: 1613-4087. DOI: https://doi.org/10.1515/commun-2019-0188. URL: https://www.degruyter.com/document/doi/10.1515/commun-2019-0188/html.

64. Dave D'Alessio and Mike Allen. "Media Bias in Presidential Elections: A Meta-Analysis". In: *Journal of Communication* 50.4 (Dec. 2000), pp. 133–156. DOI: https://doi.org/10.1111/j.1460-2466.2000.tb02866.x. URL: http://doi.wiley.com/10.1111/j.1460-2466.2000.tb02866.x.

65. Paul D'Angelo and Jim A Kuypers. *Doing news framing analysis: Empirical and theoretical perspectives*. Routledge, 2010.

66. Michal Danilak. langdetect. 2020. URL: https://github.com/Mimino666/langdetect (visited on 03/20/2021).

67. Amitava Das, Sivaji Bandyaopadhyay, and Björn Gambäck. "The 5W Structure for Sentiment Summarization-Visualization-Tracking". In: Lecture Notes in Computer Science (including subseries Lecture Notes in Artificial Intelligence and Lecture Notes in Bioinformatics). 2012, pp. 540–555. ISBN: 9783642286032. DOI: https://doi.org/10.1007/978-3-642-28604-9_44. URL: http://link.springer.com/10.1007/978-3-642-28604-944.

68. Murray S. Davis and Erving Goffman. "Frame Analysis: An Essay on the Organization of Experience." In: *Contemporary Sociology* 4.6 (Nov. 1975), p. 599. ISSN: 00943061. DOI: https://doi.org/10.2307/2064021. URL: http://www.jstor.org/stable/2064021?origin=crossref.

69. Claes H De Vreese. "News framing: Theory and typology". In: *Information design journal and document design* 13.1 (2005), pp. 51–62.

70. Lizzie Dearden. *The fake refugee images that are being used to distort public opinion on asylum seekers*. Sept. 2015. URL: http://www.independent.co.uk/news/world/europe/the-fake-refugee-images-that-are-being-usedto-distort-public-opinion-on-asylum-seekers-10503703.html (visited on 02/18/2020).

71. Stefano DellaVigna and Ethan Kaplan. The Fox News Effect: Media Bias and Voting. Tech. rep. 3. Cambridge, MA: National Bureau of Economic Research, Apr. 2006, pp. 1187–1234. DOI: https://doi.org/10.3386/w12169. URL: http://www.nber.org/papers/w12169.pdf.

72. P. M. DeMarzo, Dimitri Vayanos, and Jeffrey Zwiebel. "Persuasion Bias, Social Influence, and Unidimensional Opinions". In: *The Quarterly Journal of Economics* 118.3 (Aug. 2003), pp. 909-968. ISSN: 0033-5533. DOI: 10.1162/00335530360698469. URL: https://doi.org/10.1162/00335530360698469.

73. Jacob Devlin et al. "BERT: Pre-training of Deep Bidirectional Transformers for Language Understanding". In: *Proceedings of the 2019 Conference of the North*. Stroudsburg, PA, USA: Association for Computational Linguistics, 2019, pp. 4171–4186. DOI: https://doi.org/10.18653/v1/N19-1423. arXiv: 1810.04805. URL: http://aclweb.org/anthology/N19-1423.

74. Jeremy Diamond and Kevin Liptak. A year after 'Little Rocket Man' US-NK relations face uncertain path. 2018. URL: https://edition.cnn.com/2018/09/24/politics/north-korea-little-rocket-man-donald-trump/index.html (visited on 01/15/2021).

75. James N Druckman and Michael Parkin. "The impact of media bias: How editorial slant affects voters". In: *Journal of Politics* 67.4 (2005), pp. 1030–1049.

76. Chunning Du et al. "Adversarial and Domain-Aware BERT for Cross- Domain Sentiment Analysis". In: *Proceedings of the 58th Annual Meeting of the Association for Computational Linguistics*. Stroudsburg, PA, USA: Association for Computational Linguistics, 2020, pp. 4019–4028. DOI: https://doi.org/10.18653/v1/2020.acl-main.370. URL: https://www.aclweb.org/anthology/2020.aclmain.370.

77. Jonas Ehrhardt et al. "Omission of Information: Identifying Political Slant via an Analysis of Co-occurring Entities". In: *156h International Symposium of Information Science* (ISI 2021). Glückstadt,Germany: VerlagWerner Hülsbusch, 2021, pp. 80–93. URL: https://epub.uni-regensburg.de/44939/.

78. Erick Elejalde, Leo Ferres, and Eelco Herder. "On the nature of real and perceived bias in the mainstream media". In: *PLOS ONE* 13.3 (Mar. 2018). Ed. by Dante R. Chialvo, pp. 1–28. ISSN: 1932-6203. DOI: https://doi.org/10.1371/journal.pone.0193765. URL: https://dx.plos.org/10.1371/journal.pone.0193765.

79. RobertMEntman. "Framing: Toward Clarification of a Fractured Paradigm". In: *Journal of Communication* 43.4 (Dec. 1993), pp. 51–58. ISSN: 0021-9916. DOI: https://doi.org/10.1111/j.1460-2466.1993.tb01304.x. URL: https://academic.oup.com/joc/article/43/4/51-58/4160153.

80. Robert M. Entman. "Framing Bias: Media in the Distribution of Power". In: *Journal of Communication* 57.1 (Mar. 2007), pp. 163–173. ISSN: 00219916. DOI: https://doi.org/10.1111/j.1460-2466.2006.00336.x. URL: https://academic.oup.com/joc/article/57/1/163-173/4102665.

81. Frank Esser. "Editorial Structures and Work Principles in British and German Newsrooms". In: *European Journal of Communication* 13.3 (Sept. 1998), pp. 375–405. ISSN: 0267-3231. DOI: https://doi.org/10.1177/0267323198013003004. arXiv: 0803973233. URL: http://journals.sagepub.com/doi/10.1177/0267323198013003004.

82. Frank Esser, Carsten Reinemann, and David Fan. "Spin Doctors in theUnited States, Great Britain, and Germany Metacommunication about Media Manipulation". In: *The Harvard International Journal of Press/Politics* 6.1 (2001), pp. 16–45.

83. James Estrin. The Real Story About the Wrong Photos in #BringBackOur-Girls. May 2014. URL: http://lens.blogs.nytimes.com/2014/05/08/thereal-story-about-the-wrong-photos-in-bringbackourgirls/ (visited on 02/18/2020).

84. David Kirk Evans, Judith L. Klavans, and Kathleen R.McKeown. "Columbia Newsblaster". In: *Demonstration Papers at HLT-NAACL 2004 on XX - HLTNAACL '04*. Morristown, NJ,

USA: Association for Computational Linguistics, 2004, pp. 1–4. DOI: https://doi.org/10.3115/1614025.1614026. URL: http://portal.acm.org/citation.cfm?doid=1614025.1614026.

85. Facebook. Company Info. 2021. URL: http://web.archive.org/web/20210210223947/ https://about.fb.com/company-info/ (visited on 02/12/2021).

86. Lukas Feick, Karsten Donnay, and Katherine T. McCabe. "The Subconscious Effect of Subtle Media Bias on Perceptions of Terrorism". In: *American Politics Research* 49.3 (May 2021), pp. 313–318. ISSN: 1532-673X. DOI: https://doi.org/10.1177/1532673X20972105. URL: http://journals.sagepub.com/doi/10.1177/1532673X20972105.

87. Charles J. Fillmore and Collin F. Baker. "Frame semantics for text understanding". In: *Proceedings of NAACLWordNet and Other Lexical Resources Workshop*. Pittsburgh, US, 2001, pp. 1–6. URL: https://citeseerx.ist.psu.edu/viewdoc/download?doi=10.1.1.469.9423&rep=rep1&type=pdf.

88. Jenny Rose Finkel, Trond Grenager, and Christopher Manning. "Incorporating non-local information into information extraction systems by Gibbs sampling". In: *Proceedings of the 43rd Annual Meeting on Association for Computational Linguistics - ACL '05*. Association for Computational Linguistics. Morristown, NJ, USA: Association for Computational Linguistics, 2005, pp. 363–370. DOI: https://doi.org/10.3115/1219840.1219885. URL: http://portal.acm.org/citation.cfm?doid=1219840.1219885.

89. Tomáš Foltýnek, Norman Meuschke, and Bela Gipp. "Academic Plagiarism Detection". In: *ACM Computing Surveys* 52.6 (Jan. 2020), pp. 1–42. ISSN: 0360-0300. DOI: https://doi.org/10.1145/3345317. URL: https://dl.acm.org/doi/10.1145/3345317.

90. James Franklin. "The elements of statistical learning: data mining, inference and prediction". In: *The Mathematical Intelligencer* 27.2 (Mar. 2005), pp. 83–85. ISSN: 0343-6993. DOI: https://doi.org/10.1007/BF02985802. URL: http://link.springer.com/10.1007/BF02985802.

91. Brendan Frey and Delbert Dueck. "Clustering by Passing Messages Between Data Points". In: *Science* 315.5814 (Feb. 2007), pp. 972–976. DOI: https://doi.org/10.1126/science.1136800.

92. Dieter Frey. "Recent research on selective exposure to information". In: *Advances in experimental social psychology* 19 (1986), pp. 41–80. URL: https://doi.org/10.1016/S0065-2601(08)60212-9.

93. Mingkun Gao et al. "To Label or Not to Label: The Effect of Stance and Credibility Labels on Readers' Selection and Perception of News Articles". In: *Proceedings of the ACMon Human-Computer Interaction* 2.CSCW (Nov. 2018), pp. 1–16. ISSN: 2573-0142. DOI: https://doi.org/10.1145/3274324. URL: https://dl.acm.org/doi/10.1145/3274324.

94. Zhengjie Gao et al. "Target-Dependent Sentiment Classification With BERT". In: *IEEE Access* 7 (2019), pp. 154290–154299. ISSN: 2169-3536. DOI: https://doi.org/10.1109/ACCESS.2019.2946594. URL: https://ieeexplore.ieee.org/document/8864964/.

95. Robert Garner, Peter Ferdinand, and Stephanie Lawson. Introduction to politics. Oxford, United Kingdom: Oxford University Press, 2012, pp. 1–525. ISBN: 978-0-19-960572-9.

96. Lukas Gebhard and Felix Hamborg. "The POLUSA Dataset: 0.9M Political News Articles Balanced by Time and Outlet Popularity". In: *Proceedings of theACM/IEEE Joint Conference on Digital Libraries in 2020*. NewYork, NY, USA: ACM, Aug. 2020, pp. 467–468. ISBN: 9781450375856. DOI: https://doi.org/10.1145/3383583.3398567. URL: https://dl.acm.org/doi/10.1145/3383583.3398567.

97. Matthew Gentzkow, Edward Glaeser, and Claudia Goldin. The Rise of the Fourth Estate: How Newspapers Became Informative and Why It Mattered. Tech. rep. Cambridge, MA: National Bureau of Economic Research, Sept. 2004, pp. 187–230. DOI: https://doi.org/10.3386/w10791. URL: http://www.nber.org/papers/w10791.pdf.

98. Matthew Gentzkow and Jesse M Shapiro. "What drives media slant? Evidence from US daily newspapers". In: *Econometrica* 78.1 (2010), pp. 35–71. URL: https://web.stanford.edu/~gentzkow/research/biasmeas.pdf.

99. Matthew Gentzkow and Jesse M. Shapiro. "Media Bias and Reputation". In: *Journal of Political Economy* 114.2 (Apr. 2006), pp. 280–316. ISSN: 0022-3808. DOI: 10.1086/499414. URL: https://doi.org/10.1086/499414.

100. Alan S Gerber, Dean Karlan, and Daniel Bergan. "Does the media matter? A field experiment measuring the effect of newspapers on voting behavior and political opinions". In: *American Economic Journal: Applied Economics* 1.2 (2009), pp. 35–52.

101. Ran Geva. *article-date-extractor*. 2018. URL: https://github.com/Webhose/article-date-extractor (visited on 02/17/2021).

102. Deepanway Ghosal et al. "KinGDOM: Knowledge-Guided DOMainAdaptation for Sentiment Analysis". In: *Proceedings of the 58th Annual Meeting of the Association for Computational Linguistics*. Stroudsburg, PA, USA: Association for Computational Linguistics, 2020, pp. 3198–3210. DOI: https://doi.org/10.18653/v1/2020.acl-main.292. URL: https://www.aclweb.org/anthology/2020.aclmain.292.

103. Martin Gilens and Craig Hertzman. "Corporate ownership and news bias: Newspaper coverage of the 1996 Telecommunications Act". In: *The Journal of Politics* 62.02 (2000), pp. 369–386.

104. Roger Giner-Sorolla and Shelly Chaiken. "The Causes of Hostile Media Judgments". In: *Journal of Experimental Social Psychology* 30.2 (Mar. 1994), pp. 165–180. ISSN: 00221031. DOI: https://doi.org/10.1006/jesp.1994.1008. URL: https://linkinghub.elsevier.com/retrieve/pii/S0022103184710080.

105. Bela Gipp. Citation-based Plagiarism Detection. Wiesbaden: Springer FachmedienWiesbaden, 2014. ISBN: 978-3-658-06393-1. DOI: https://doi.org/10.1007/978-3-658-06394-8. URL: http://link.springer.com/10.1007/978-3-658-06394-8.

106. Bela Gipp, Norman Meuschke, and Corinna Breitinger. "Citation-based plagiarism detection: Practicability on a large-scale scientific corpus". In: *Journal of the Association for Information Science and Technology* 65.8 (Aug. 2014), pp. 1527–1540. ISSN: 23301635. DOI: https://doi.org/10.1002/asi.23228. arXiv: 0803.1716. URL: http://doi.wiley.com/10.1002/asi.23228.

107. Bela Gipp, Adriana Taylor, and Jöran Beel. 'Link Proximity Analysis - ClusteringWebsites by Examining Link Proximity". In: 2010, pp. 449–452. DOI: https://doi.org/10.1007/978-3-642-15464-5_54. URL: http://link.springer.com/10.1007/978-3-642-15464-554.

108. Roxana Girju. "Automatic detection of causal relations for Question Answering". In: *Proceedings of the ACL 2003 workshop on Multilingual summarization and question answering -*. Vol. 12. Morristown, NJ, USA: Association for Computational Linguistics, 2003, pp. 76–83. DOI: https://doi.org/10.3115/1119312.1119322. URL: http://portal.acm.org/citation.cfm?doid=1119312.1119322.

109. Xavier Glorot and Yoshua Bengio. "Understanding the difficulty of training deep feedforward neural networks". In: *Journal of Machine Learning Research*. 2010, pp. 249–256.

110. Namrata Godbole, Manja Srinivasaiah, and Steven Skiena. "Large-Scale Sentiment Analysis for News and Blogs". In: *Proceedings of the International Conference onWeblogs and Social Media (ICWSM) 7* (2007), pp. 219–222.

111. Jennifer Golbeck and Derek Hansen. "Computing political preference among twitter followers". In: *Proceedings of the SIGCHI Conference on Human Factors in Computing Systems*. ACM. 2011, pp. 1105–1108.

112. Jonathan Goldsmith. wikipedia - Python Package. 2014. URL: https://github.com/goldsmith/Wikipedia (visited on 04/19/2021).

113. Ian Goodfellow, Yoshua Bengio, and Aaron Courville. *Deep Learning*. MIT Press, 2016. URL: http://www.deeplearningbook.org.

114. Gravity Labs. Goose - Article Extractor. 2021. URL: https://github.com/GravityLabs/goose (visited on 02/17/2021).

115. Derek Greene and Pádraig Cunningham. "Practical solutions to the problem of diagonal dominance in kernel document clustering". In: *Proceedings of the 23rd international conference on Machine learning - ICML '06*. New York, New York, USA: ACM Press, 2006, pp. 377–384. ISBN: 1595933832. DOI: https://doi.org/10.1145/1143844.1143892. URL: http://portal.acm.org/citation.cfm?doid=1143844.1143892.

116. Gregory Grefenstette et al. "Coupling Niche Browsers and Affect Analysis for an Opinion Mining Application". In: *Coupling Approaches, Coupling Media and Coupling Languages for Information Retrieval*. Vaucluse, France: Le Centre de Hautes Etudes Internationales

D'Informatique Documentaire, 2004, pp. 186–194. URL: https://dl.acm.org/doi/abs/10.5555/2816272.2816290.

117. Tim Groseclose and Jeffrey Milyo. "A Measure of Media Bias". In: The Quarterly Journal of Economics 120.4 (Nov. 2005), pp. 1191–1237. ISSN: 0033-5533. DOI: https://doi.org/10.1162/003355305775097542. URL: http://dx.doi.org/10.1162/003355305775097542.

118. Nicole Gruber and Alfred Jockisch. "Are GRU Cells More Specific and LSTM Cells More Sensitive in Motive Classification of Text?" In: Frontiers in Artificial Intelligence 3 (June 2020). ISSN: 2624-8212. DOI: https://doi.org/10.3389/frai.2020.00040. URL: https://www.frontiersin.org/article/10.3389/frai.2020.00040/full.

119. Jeff Gruenewald, Jesenia Pizarro, and Steven M. Chermak. "Race, gender, and the newsworthiness of homicide incidents". In: Journal of Criminal Justice 37.3 (May 2009), pp. 262–272. ISSN: 00472352. DOI: https://doi.org/10.1016/j.jcrimjus.2009.04.006. URL: https://linkinghub.elsevier.com/retrieve/pii/S0047235209000440.

120. Joachim W H Haes. "September 11 in Germany and the United States: Reporting, reception, and interpretation". In: Crisis Communications: Lessons from September 11 (2003), pp. 125–132.

121. Jens Hainmueller, Dominik Hangartner, and Teppei Yamamoto. "Validating vignette and conjoint survey experiments against real-world behavior". In: Proceedings of the National Academy of Sciences 112.8 (Feb. 2015), pp. 2395–2400. ISSN: 0027-8424. DOI: https://doi.org/10.1073/pnas.1416587112. URL: http://www.pnas.org/lookup/doi/10.1073/pnas.1416587112.

122. Jens Hainmueller, Daniel J. Hopkins, and Teppei Yamamoto. "Causal Inference in Conjoint Analysis: Understanding Multidimensional Choices via Stated Preference Experiments". In: Political Analysis 22.1 (Jan. 2014), pp. 1–30. ISSN: 1047-1987. DOI: https://doi.org/10.1093/pan/mpt024. URL: https://www.cambridge.org/core/product/identifier/S1047198700013589/type/journalarticle.

123. Felix Hamborg. "Media Bias, the Social Sciences, and NLP: Automating Frame Analyses to Identify Bias byWord Choice and Labeling". In: Proceedings of the 58th Annual Meeting of the Association for Computational Linguistics: Student Research Workshop. Stroudsburg, PA, USA: Association for Computational Linguistics, 2020, pp. 79–87. DOI: https://doi.org/10.18653/v1/2020.aclsrw.12. URL: https://www.aclweb.org/anthology/2020.acl-srw.12.

124. Felix Hamborg, Corinna Breitinger, and Bela Gipp. "Giveme5W1H: A Universal System for Extracting Main Events from News Articles". In: Proceedings of the 13th ACM Conference on Recommender Systems, 7th International Workshop on News Recommendation and Analytics (INRA 2019). Copenhagen, Denmark, 2019, pp. 1–8.

125. Felix Hamborg and Karsten Donnay. "NewsMTSC: A Dataset for (Multi-)Target-dependent Sentiment Classification in Political News Articles". In: Proceedings of the 2021 Conference of the European Chapter of the Association for Computational Linguistics (EACL). Association for Computational Linguistics, 2021, pp. 1–13.

126. Felix Hamborg, Karsten Donnay, and Bela Gipp. "Automated identification of media bias in news articles: an interdisciplinary literature review". In: International Journal on Digital Libraries 20.4 (Dec. 2019), pp. 391–415. ISSN: 1432-5012. DOI: https://doi.org/10.1007/s00799-018-0261-y. URL: http://link.springer.com/10.1007/s00799-018-0261-y.

127. Felix Hamborg, Karsten Donnay, and Bela Gipp. "TowardsTarget-Dependent Sentiment Classification in News Articles". In: Proceedings of the 16th iConference. Ed. by Katharina Toeppe et al. Be.ing, China (Virtual Event): Springer, 2021, pp. 156–166. DOI: https://doi.org/10.1007/978-3-030-71305-8_12. URL: http://link.springer.com/10.1007/978-3-030-71305-8_12.

128. Felix Hamborg, Norman Meuschke, and Bela Gipp. "Bias-aware news analysis using matrix-based news aggregation". In: International Journal on Digital Libraries 21.2 (June 2020), pp. 129–147. ISSN: 1432–5012. DOI: https://doi.org/10.1007/s00799-018-0239-9. URL: http://link.springer.com/10.1007/s00799-018-0239-9.

129. Felix Hamborg, Norman Meuschke, and Bela Gipp. "Matrix-Based News Aggregation: Exploring Different Newsctives". In: *2017 ACM/IEEE Joint Conference on Digital Libraries (JCDL)*. IEEE, June 2017, pp. 1–10. ISBN: 978-1-5386-3861-3. DOI: https://doi.org/10.1109/ JCDL.2017.7991561. URL: http://ieeexplore.ieee.org/document/7991561/.

130. Felix Hamborg, Anastasia Zhukova, and Bela Gipp. "Automated Identification of Media Bias by Word Choice and Labeling in News Articles". In: *2019 ACM/IEEE Joint Conference on Digital Libraries (JCDL)*. Champaign, IL, USA: IEEE, June 2019, pp. 196–205. ISBN: 978-1-7281-1547-4. DOI: https://doi.org/10.1109/JCDL.2019.00036. URL: https://ieeexplore.ieee. org/document/8791197/.

131. Felix Hamborg, Anastasia Zhukova, and Bela Gipp. "Illegal Aliens or Undocumented Immigrants? Towards the Automated Identification of Bias by Word Choice and Labeling". In: *Information in Contemporary Society: iConference 2019*. Washington, DC, USA: Springer International Publishing, 2019, pp. 179–187. DOI: https://doi.org/10.1007/978-3-030-15742-5_17. URL: http://link.springer.com/10.1007/978-3-030-15742-517.

132. Felix Hamborg et al. "Extraction of Main Event Descriptors from News Articles by Answering the Journalistic Five W and One H Questions". In: *Proceedings of the 18th ACM/IEEE on Joint Conference on Digital Libraries*. New York, NY, USA: ACM, May 2018, pp. 339–340. ISBN: 9781450351782. DOI: https://doi.org/10.1145/3197026.3203899. URL: https://dl.acm.org/doi/10.1145/3197026.3203899.

133. Felix Hamborg et al. "Giveme5W: Main Event Retrieval from News Articles by Extraction of the Five Journalistic W Questions". In: *Proceedings of the iConference 2018: Transforming Digital Worlds*. Ed. by Gobinda Chowdhury. Sheffield, UK: Springer International Publishing, 2018, pp. 356–366. DOI: https://doi.org/10.1007/978-3-319-78105-1_39. URL: http://link. springer.com/10.1007/978-3-319-78105-1_39.

134. Felix Hamborg et al. "How to Effectively Identify and Communicate Person- Targeting Media Bias in Daily News Consumption?" In: *Proceedings of the 15th ACM Conference on Recommender Systems, 9th International Workshop on News Recommendation and Analytics (INRA 2021)*. Amsterdam, Netherlands: Association for Computing Machinery, 2021, pp. 1–11.

135. Felix Hamborg et al. "Identification and Analysis of Media Bias in News Articles". In: *15th International Symposium of Information Science (ISI 2017)*. Berlin, Germany: Verlag Werner Hülsbusch, 2017, pp. 224–236. ISBN: 978-3-86488-117-6.

136. Felix Hamborg et al. "news-please: a Generic News Crawler and Extractor". In: *15th International Symposium of Information Science (ISI 2017)*. Berlin, Germany: Verlag Werner Hülsbusch, 2017, pp. 218–223. ISBN: 978-3-86488-117-6.

137. Felix Hamborg et al. "Newsalyze: Effective Communication of Person- Targeting Biases in News Articles". In: *2021 ACM/IEEE Joint Conference on Digital Libraries (JCDL)*. Los Alamitos, CA, USA: IEEE, 2021, pp. 130–139. DOI: https://doi.org/10.1109/JCDL52503. 2021.00025. URL: https://doi.ieeecomputersociety.org/10.1109/JCDL52503.2021.00025.

138. Felix Hamborg et al. "Newsalyze: Enabling News Consumers to Understand Media Bias". In: *Proceedings of the ACM/IEEE Joint Conference on Digital Libraries in 2020*. New York, NY, USA: ACM, Aug. 2020, pp. 455–456. ISBN: 9781450375856. DOI: https://doi.org/10.1145/ 3383583.3398561. URL: https://dl.acm.org/doi/10.1145/3383583.3398561.

139. Felix Hamborg et al. "NewsDeps: Visualizing the Origin of Information in News Articles". In: *Wahrheit und Fake im postfaktisch-digitalen Zeitalter*. Ed. by Peter Klimczak and Thomas Zoglauer. Springer Vieweg, 2021, pp. 151–166. ISBN: 978-3-658-32957-0. DOI: https://doi. org/10.1007/978-3-658-32957-0.

140. Mark Hanna. "Keywords in News and Journalism Studies". In: Journalism Studies 15.1 (Jan. 2014), pp. 118–119. ISSN: 1461-670X. DOI: https://doi.org/10.1080/1461670X.2012.712759. URL: http://www.tandfonline.com/doi/abs/10. 1080/1461670X.2012.712759.

141. Tony Harcup and Deirdre O'neill. "What is news? Galtung and Ruge revisited". In: *Journalism studies* 2.2 (2001), pp. 261–280. URL: https://www.tandfonline.com/doi/10.1080/ 14616700118449.

142. Martie G. Haselton, Daniel Nettle, and Damian R. Murray. "The Evolution of Cognitive Bias". In: *The Handbook of Evolutionary Psychology*. Hoboken, NJ, USA: John Wiley & Sons, Inc., Nov. 2015, pp. 1–20. DOI: https://doi.org/10.1002/9781119125563.evpsych241. URL: http://doi.wiley.com/10.1002/9781119125563.evpsych241.

143. Laura Hasler, Constantin Orasan, and Karin Naumann. "NPs for Events: Experiments in Coreference Annotation". In: *Proceedings of the Fifth International Conference on Language Resources and Evaluation (LREC)*. 2006, pp. 1167–1172. URL: http://www.lrec-conf.org/proceedings/lrec2006/pdf/539_pdf.pdf.

144. Andrew F. Hayes and Klaus Krippendorff. "Answering the Call for a Standard Reliability Measure for Coding Data". In: *Communication Methods and Measures* 1.1 (Apr. 2007), pp. 77–89. ISSN: 1931-2458. DOI: https://doi.org/10.1080/19312450709336664. URL: http://www.tandfonline.com/doi/abs/10.1080/19312450709336664.

145. David Heise. "The semantic differential and attitude research". In: *Attitude measurement* 4 (1970), pp. 235–253.

146. Edward S Herman. "The propaganda model:Aretrospective". In: *Journalism Studies* 1.1 (2000), pp. 101–112. DOI: https://doi.org/10.1080/146167000361195.

147. Edward S Herman and Noam Chomsky. *Manufacturing consent: The political economy of the mass media*. Random House, 2010.

148. John R. Hibbing, Kevin B. Smith, and John R. Alford. "Differences in negativity bias underlie variations in political ideology". In: *Behavioral and Brain Sciences* 37.3 (June 2014), pp. 297–307. ISSN: 0140-525X. DOI: https://doi.org/10.1017/S0140525X13001192. URL: https://www.cambridge.org/core/product/identifier/S0140525X13001192/type/journal%7B%5C_%7Darticle.

149. Timothy C Hoad and Justin Zobel. "Methods for identifying versioned and plagiarized documents". In: *Journal of the American society for information science and technology* 54.3 (2003), pp. 203–215.

150. Johannes Hoffart et al. "Robust Disambiguation of Named Entities in Text". In: *Proceedings of the 2011 Conference on Empirical Methods in Natural Language Processing*. Edinburgh, Scotland, UK: Association for Computational Linguistics, 2011, pp. 782–792. URL: https://www.aclweb.org/anthology/D11-1072.

151. Marjan Hosseinia, Eduard Dragut, and Arjun Mukherjee. "Stance Prediction for Contemporary Issues: Data and Experiments". In: *Proceedings of the Eighth International Workshop on Natural Language Processing for Social Media*. Stroudsburg, PA, USA: Association for Computational Linguistics, 2020, pp. 32–40. DOI: https://doi.org/10.18653/v1/2020.socialnlp-1.5. URL: https://www.aclweb.org/anthology/2020.socialnlp-1.5.

152. George Hripcsak. "Agreement, the F-Measure, andReliability in Information Retrieval". In: *Journal of the American Medical Informatics Association* 12.3 (Jan. 2005), pp. 296–298. ISSN: 1067-5027. DOI: https://doi.org/10.1197/jamia.M1733. URL: https://academic.oup.com/jamia/article-lookup/doi/10.1197/jamia.M1733.

153. Minqing Hu and Bing Liu. "Mining and summarizing customer reviews". In: *Proceedings of the 2004 ACM SIGKDD international conference on Knowledge discovery and data mining - KDD '04*. New York, New York, USA: ACM Press, 2004, p. 168. DOI: https://doi.org/10.1145/1014052.1014073. URL: http://portal.acm.org/citation.cfm?doid=1014052.1014073.

154. Christoph Hube, Besnik Fetahu, and Ujwal Gadiraju. "Understanding and Mitigating Worker Biases in the Crowdsourced Collection of Subjective Judgments". In: *Proceedings of the 2019 CHI Conference on Human Factors in Computing Systems*. New York, NY, USA: ACM, May 2019, pp. 1–12. ISBN: 9781450359702. DOI: https://doi.org/10.1145/3290605.3300637. URL: https://dl.acm.org/doi/10.1145/3290605.3300637.

155. John Edward Hunter, Frank L Schmidt, and Gregg B Jackson. *Meta-analysis: Cumulating research findings across studies*. Vol. 4. Sage Publications, Inc, 1982.

156. Ichiro Ide et al. "TrackThem: Exploring a large-scale news video archive by tracking human relations". In: *Lecture Notes in Computer Science (including subseries Lecture Notes in Artificial Intelligence and Lecture Notes in Bioinformatics)*. Vol. 3689 LNCS. 2005, pp. 510–515. ISBN: 3540291865. DOI: https://doi.org/10.1007/11562382_42.

157. Takahiro Ikeda, Akitoshi Okumura, and Kazunori Muraki. "Information classification and navigation based on 5W1H of the target information". In: *Proceedings of the 36th annual meeting on Association for Computational Linguistics* -. Vol. 1. Morristown, NJ, USA: Association for Computational Linguistics, 1998, p. 571. DOI: https://doi.org/10.3115/980845. 980940. URL: http://portal.acm.org/citation.cfm?doid=980845.980940.

158. Intel AI Lab. *NLP Architect*. 2018. *doi*: 10.5281/zenodo.1477518. URL: https://doi.org/10. 5281/zenodo.1477518 (visited on 04/15/2021).

159. Shanto Iyengar. *Is anyone responsible? How television frames political issues*. University of Chicago Press, 1994.

160. Anil K Jain and Sushil Bhattacharjee. "Text segmentation using Gabor filters for automatic document processing". In: *Machine Vision and Applications* 5.3 (1992), pp. 169–184. URL: https://link.springer.com/article/10.1007/BF02626996.

161. Qingnan Jiang et al. "A Challenge Dataset and Effective Models for Aspect-Based Sentiment Analysis". In: *Proceedings of the 2019 Conference on Empirical Methods in Natural Language Processing and the 9th International Joint Conference on Natural Language Processing (EMNLP-IJCNLP)*. Stroudsburg, PA, USA: Association for Computational Linguistics, 2019, pp. 6279–6284. DOI: https://doi.org/10.18653/v1/D19-1654. URL: https:// www.aclweb.org/anthology/D19-1654.

162. S.E. Jørgensen. "Model Selection and Multimodel Inference". In: *Ecological Modelling* 172.1 (Feb. 2004), pp. 96–97. ISSN: 03043800. DOI: https://doi.org/10.1016/j.ecolmodel.2003.11. 004. URL: https://linkinghub.elsevier.com/retrieve/pii/S0304380003004526.

163. Brihi Joshi et al. "The Devil is in the Details: Evaluating Limitations of Transformer-based Methods for Granular Tasks". In: (Nov. 2020). arXiv: 2011.01196. URL: http://arxiv.org/abs/ 2011.01196.

164. Silvia Julinda, Christoph Boden, and Alan Akbik. *Extracting a Repository of Events and Event References fromNews Clusters*. Dublin, Ireland,Aug. 2014. DOI: https://doi.org/10. 3115/v1/W14-4503. URL: https://www.aclweb.org/anthology/W14-4503.

165. Daniel Jurafsky and James H Martin. "Speech and language processing". In: 710 (2000), pp. 1–1032. URL: https://web.stanford.edu/~jurafsky/slp3/.

166. Daniel Kahneman and Amos Tversky. "Choices, values, and frames." In: *American Psychologist* 39.4 (1984), pp. 341–350. ISSN: 0003-066X. DOI: https://doi.org/10.1037/0003-066X. 39.4.341. URL: http://content.apa.org/journals/amp/39/4/341.

167. Mesut Kaya, Guven Fidan, and Ismail H Toroslu. "Sentiment Analysis of Turkish Political News". In: *2012 IEEE/WIC/ACM International Conferences on Web Intelligence and Intelligent Agent Technology*. IEEE Computer Society. IEEE, Dec. 2012, pp. 174–180. ISBN: 978-1-4673-6057-9. DOI: https://doi.org/10.1109/WI-IAT.2012.115. URL: http://ieeexplore. ieee.org/document/6511881/.

168. Jaana Kekäläinen and Kalervo Järvelin. "Using graded relevance assessments in IR evaluation". In: *Journal of the American Society for Information Science and Technology* 53.13 (Nov. 2002), pp. 1120–1129. ISSN: 15322882. DOI: https://doi.org/10.1002/asi.10137. URL: http://doi.wiley.com/10.1002/asi.10137.

169. Keith Kenney and Chris Simpson. "Was coverage of the 1988 presidential race by Washington's two major dailies biased?" In: *Journalism & Mass Communication Quarterly* 70.2 (1993), pp. 345–355. DOI: https://doi.org/10.1177/107769909307000210.

170. Kian Kenyon-Dean, Jackie Chi Kit Cheung, and Doina Precup. "Resolving Event Coreference with Supervised Representation Learning and Clustering- OrientedRegularization". In: *Proceedings of the Seventh Joint Conference on Lexical and Computational Semantics*. Stroudsburg, PA, USA: Association for Computational Linguistics, 2018, pp. 1–10. DOI: https://doi.org/10.18653/v1/S18-2001. URL: http://aclweb.org/anthology/S18-2001.

171. Hans Mathias Kepplinger. "Visual biases in television campaign coverage". In: *Communication Research* 9.3 (1982), pp. 432–446. DOI: https://doi.org/10.1177/009365082009003005.

172. Jean S Kerrick. "News pictures, captions and the point of resolution". In: *Journalism & Mass Communication Quarterly* 36.2 (1959), pp. 183–188. DOI: https://doi.org/10.1177/ 107769905903600207.

173. Jean S. Kerrick. "The Influence of Captions on Picture Interpretation". In: *Journalism Quarterly* 32.2 (June 1955), pp. 177–182. ISSN: 0022-5533. DOI: https://doi.org/10.1177/107769905503200205. URL: http://journals.sagepub.com/doi/10.1177/107769905503200205.

174. Aliakbar Keshtkaran, Siti Sophiayati Yuhaniz, and Suhaimi Ibrahim. "An overview of cross-document coreference resolution". In: *2017 International Conference on Computer and Drone Applications (IConDA)*. IEEE, Nov. 2017, pp. 43–48. ISBN: 978-1-5386-0765-7. DOI: https://doi.org/10.1109/ICONDA.2017.8270397. URL: http://ieeexplore.ieee.org/document/8270397/.

175. Masayu Leylia Khodra. "Event extraction on Indonesian news article using multiclass categorization". In: *2015 2nd International Conference on Advanced Informatics: Concepts, Theory and Applications (ICAICTA)*. IEEE, Aug. 2015, pp. 1–5. ISBN: 978-1-4673-8142-0. DOI: https://doi.org/10.1109/ICAICTA.2015.7335365. URL: http://ieeexplore.ieee.org/document/7335365/.

176. C. S. G. Khoo et al. "Automatic Extraction of Cause-Effect Information from Newspaper Text Without Knowledge-based Inferencing". In: *Literary and Linguistic Computing* 13.4 (Dec. 1998), pp. 177–186. ISSN: 0268-1145. doi: https://doi.org/10.1093/llc/13.4.177. URL: https://academic.oup.com/dsh/articlelookup/doi/10.1093/llc/13.4.177.

177. Christopher S G Khoo. "Automatic identification of causal relations in text and their use for improving precision in information retrieval". PhD thesis. 1995. URL: https://surface.syr.edu/it_etd/36/.

178. Johannes Kiesel et al. "SemEval-2019 Task 4: Hyperpartisan News Detection". In: *Proceedings of the 13th International Workshop on Semantic Evaluation*. Stroudsburg, PA, USA: Association for Computational Linguistics, 2019, pp. 829–839. DOI: https://doi.org/10.18653/v1/S19-2145. URL: https://www.aclweb.org/anthology/S19-2145.

179. JongWook Kim, K Selçuk Candan, and Junichi Tatemura. "Efficient overlap and content reuse detection in blogs and online news articles". In: *Proceedings of the 18th international conference on World wide web - WWW '09*. 0735014. New York, New York, USA: ACM Press, 2009, p. 81. ISBN: 9781605584874. DOI: https://doi.org/10.1145/1526709.1526721. URL: http://portal.acm.org/citation.cfm?doid=1526709.1526721.

180. Sung-Hee Kim, Hyokun Yun, and Ji Soo Yi. "How to filter out random clickers in a crowdsourcing-based study?" In: *Proceedings of the 2012 BELIV Workshop on Beyond Time and Errors - Novel Evaluation Methods for Visualization - BELIV '12*. New York, New York, USA: ACM Press, 2012, pp. 1–7. ISBN: 9781450317917. DOI: https://doi.org/10.1145/2442576.2442591. URL: http://dl.acm.org/citation.cfm?doid=2442576.2442591.

181. Diederik P. Kingma and Jimmy Ba. "Adam: A Method for Stochastic Optimization". In: *3rd International Conference on Learning Representations, ICLR 2015, San Diego, CA, USA, May 7–9, 2015*, Conference Track Proceedings. 2015. URL: http://arxiv.org/abs/1412.6980.

182. Svetlana Kiritchenko et al. "NRC-Canada-2014: Detecting Aspects and Sentiment in Customer Reviews". In: *Proceedings of the 8th InternationalWorkshop on Semantic Evaluation (SemEval 2014)*. Dublin, Ireland: Association for Computational Linguistics, 2014, pp. 437–442. DOI: https://doi.org/10.3115/v1/s14-2076.

183. Knight Foundation. *American Views 2020: Trust, Media and Democracy*. 2020. URL: https://knightfoundation.org/reports/american-views-2020-trust-media-and-democracy/ (visited on 04/24/2021).

184. Caleb M. Koch et al. "Public debate in the media matters: evidence from the European refugee crisis". In: EPJ Data Science 9.1 (Dec. 2020), pp. 1–12. issn: 2193–1127. DOI: https://doi.org/10.1140/epjds/s13688-020-00229-8. URL: https://epjdatascience.springeropen.com/articles/10.1140/epjds/s13688-020-00229-8.

185. Christian Kohlschütter, Peter Fankhauser, and Wolfgang Nejdl. "Boilerplate detection using shallow text features". In: *Proceedings of the third ACM international conference on Web search and data mining - WSDM '10*. New York, New York, USA: ACM Press, 2010, p. 441. ISBN: 9781605588896. doi: https://doi.org/10.1145/1718487.1718542. URL: http://portal.acm.org/citation.cfm?doid=1718487.1718542.

186. Nikhil L. Kolluri and Dhiraj Murthy. "CoVerifi: A COVID-19 news verification system". In: *Online Social Networks and Media 22* (Mar. 2021), p. 100123. ISSN: 24686964. DOI: https://doi.org/10.1016/j.osnem.2021.100123. URL: https://linkinghub.elsevier.com/retrieve/pii/S2468696421000070.

187. Ha-Kyung Kong, Zhicheng Liu, and Karrie Karahalios. "Frames and Slants in Titles of Visualizations on Controversial Topics". In: *Proceedings of the 2018 CHI Conference on Human Factors in Computing Systems*. New York, NY, USA: ACM, Apr. 2018, pp. 1–12. ISBN: 9781450356206. DOI: https://doi.org/10.1145/3173574.3174012. URL: https://dl.acm.org/doi/10.1145/3173574.3174012.

188. Dimitrios Kouzis-Loukas. *Learning Scrapy*. Packt Publishing Ltd, 2016. isbn: 9781784399788.

189. Wolfgang Kreißig. *Medienvielfaltsmonitor 2020-I: Anteile der Medienangebote und Medienkonzerne am Meinungsmarkt der Medien in Deutschland*. Tech. rep. Munich, Germany: Bayerische Landeszentrale für neue Medien (BLM), 2020. URL: https://www.blm.de/files/pdf2/medienvielfaltsmonitor-2020-1.pdf.

190. Steven Kull, Clay Ramsay, and Evan Lewis. "Misperceptions, the media, and the Iraqwar". In: *Political Science Quarterly* 118.4 (2003), pp. 569–598. URL: https://onlinelibrary.wiley.com/doi/10.1002/j.1538-165X.2003.tb00406.x.

191. Ankit Kumar et al. "Ask Me Anything: Dynamic Memory Networks for Natural Language Processing". In: *arXiv* (2015). ISSN: 1938–7228. DOI: https://doi.org/10.1017/CBO9781107415324.004. arXiv: arXiv:1506.07285v1.

192. Matt J Kusner et al. "From Word Embeddings To Document Distances". In: *Proceedings of The 32nd International Conference on Machine Learning* 37 (2015), pp. 957–966. ISSN: 1938-7228.

193. Haewoon Kwak et al. "FrameAxis: Characterizing Framing Bias and Intensity with Word Embedding". In: (Feb. 2020), pp. 1–24. arXiv: 2002.08608. url: http://arxiv.org/abs/2002.08608.

194. George Lakoff. "Women, fire, and dangerous things". In: *What categories reveal about the mind* (1987).

195. J Richard Landis and Gary G Koch. "The Measurement of Observer Agreement for Categorical Data". In: *Biometrics* 33.1 (Mar. 1977), p. 159. ISSN: 0006341X. DOI: https://doi.org/10.2307/2529310. URL: https://www.jstor.org/stable/2529310?origin=crossref.

196. Valentino Larcinese, Riccardo Puglisi, and James M Snyder. "Partisan bias in economic news: Evidence on the agenda-setting behavior of US newspapers". In: *Journal of Public Economics* 95.9 (2011), pp. 1178–1189.

197. Quoc V. Le and Tomas Mikolov. "Distributed Representations of Sentences and Documents". In: *International Conference on Machine Learning - ICML* 2014 32 (May 2014). arXiv: 1405.4053. URL: http://arxiv.org/abs/1405.4053.

198. Yann Lecun, Yoshua Bengio, and Geoffrey Hinton. "Deep learning". In: *Nature* 521.7553 (2015), pp. 436–444. ISSN: 14764687. DOI: https://doihorg/10.1038/nature14539. arXiv: arXiv:1312.6184v5.

199. Heeyoung Lee et al. "Joint Entity and Event Coreference Resolution across Documents". In: *Proceedings of the 2012 Joint Conference on Empirical Methods in Natural Language Processing and Computational Natural Language Learning*. Jeju Island, Korea: Association for Computational Linguistics, 2012, pp. 489–500. URL: https://www.aclweb.org/anthology/D12-1045/.

200. Kenton Lee et al. "End-to-end Neural Coreference Resolution". In: *Proceedings of the 2017 Conference on Empirical Methods in Natural Language Processing*. Stroudsburg, PA, USA: Association for Computational Linguistics, 2017, pp. 188–197. DOI: https://doi.org/10.18653/v1/D17-1018. URL: http://aclweb.org/anthology/D17-1018.

201. Kalev Leetaru and Philip A Schrodt. "GDELT: Global Data on Events, Location and Tone, 1979-2012". In: *Annual Meeting of the International Studies Association* (2013), pp. 1–51. URL: http://data.gdeltproject.org/documentation/ISA.2013.GDELT.pdf.

202. Gaël Lejeune et al. "Multilingual event extraction for epidemic detection". In: *Artificial Intelligence in Medicine* 65.2 (Oct. 2015), pp. 131–143. ISSN: 09333657. DOI: https://doi.org/10.1016/j.artmed.2015.06.005. URL: https://linkinghub.elsevier.com/retrieve/pii/S0933365715000846.

203. Jure Leskovec, Lars Backstrom, and Jon Kleinberg. "Meme-tracking and the dynamics of the news cycle". In: *Proceedings of the 15th ACM SIGKDD international conference on Knowledge discovery and data mining*. ACM. 2009, pp. 497–506. DOI: https://doi.org/10.1145/1557019.1557077.

204. Vladimir I Levenshtein. "Binary codes capable of correcting deletions, insertions, and reversals". In: *Soviet Physics Doklady* 10.8 (1966), pp. 707–710.

205. David D Lewis et al. "RCV1: A New Benchmark Collection for Text Categorization Research". In: *The Journal of Machine Learning Research 5* (2004), pp. 361–397. URL: *https://dl.acm.org/doi/10.5555/1005332.1005345*.

206. LexisNexis. *LexisNexis Police Reports*. 2020. URL: http://web.archive.org/web/20200405053436/https://policereports.lexisnexis.com/search/search (visited on 02/12/2020).

207. Huifeng Li et al. "InfoXtract location normalization". In: *Proceedings of the HLT-NAACL 2003 workshop on Analysis of geographic references*. Vol. 1. Morristown, NJ, USA: Association for Computational Linguistics, 2003, pp. 39–44. DOI: https://doi.org/10.3115/1119394.1119400. URL: http://portal.acm.org/citation.cfm?id=1119400.

208. Xin Li et al. "A Unified Model for Opinion Target Extraction and Target Sentiment Prediction". In: *Proceedings of the AAAI Conference on Artificial Intelligence* 33 (July 2019), pp. 6714–6721. ISSN: 2374-3468. DOI: https://doi.org/10.1609/aaai.v33i01.33016714. URL: https://aaai.org/ojs/index.php/AAAI/article/view/4643.

209. Rochelle Lieber and Pavol Štekauer. *The Oxford Handbook of Compounding*. Ed. by Rochelle Lieber and Pavol Štekauer. Oxford University Press, July 2011. ISBN: 9780199695720. DOI: https://doi.org/10.1093/oxfordhb/9780199695720.001.0001. arXiv: arXiv:1011.1669v3. URL: http://www.oxfordhandbooks.com/view/10.1093/oxfordhb/9780199695720.001.0001/oxfordhb-9780199695720.

210. Sora Lim, Adam Jatowt, and Masatoshi Yoshikawa. "Towards Bias Inducing Word Detection by Linguistic Cue Analysis in News Articles". In: *DEIM Forum 2018*. 2018, pp. 1–6. URL: https://db-event.jpn.org/deim2018/data/papers/275.pdf.

211. Sora Lim et al. "Annotating and Analyzing Biased Sentences in News Articles using Crowdsourcing". In: *Proceedings of the 12th Language Resources and Evaluation Conference*. Marseille, France: European Language Resources Association, 2020, pp. 1478–1484. URL: https://www.aclweb.org/anthology/2020.lrec-1.184.

212. Bing Liu. "Sentiment Analysis and Opinion Mining". In: *Synthesis Lectures on Human Language Technologies* 5.1 (May 2012), pp. 1–167. ISSN: 1947-4040. DOI: https://doi.org/10.2200/S00416ED1V01Y201204HLT016. URL: http://www.morganclaypool.com/doi/abs/10.2200/S00416ED1V01Y201204HLT016.

213. Pengfei Liu, Shafiq Joty, and Helen Meng. "Fine-grained Opinion Mining with Recurrent Neural Networks and Word Embeddings". In: *Proceedings of the 2015 Conference on Empirical Methods inNatural Language Processing*. Stroudsburg, PA, USA: Association for Computational Linguistics, 2015, pp. 1433–1443. DOI: https://doi.org/10.18653/v1/D15-1168. URL: http://aclweb.org/anthology/D15-1168.

214. Yinhan Liu et al. "RoBERTa: A Robustly Optimized BERT Pretraining Approach". In: (July 2019). arXiv: 1907.11692. URL: http://arxiv.org/abs/1907.11692.

215. Will Lowe. "Software for content analysis-A Review". In: *Cambridge: Weatherhead Center for International Affairs and the Harvard Identity Project* (2002).

216. Jing Lu and Vincent Ng. "Event Coreference Resolution with Multi-Pass Sieves". In: *Proceedings of the Tenth International Conference on Language Resources and Evaluation (LREC 2016)*. Portoroz, Slovenia: European Language Resources Association (ELRA), 2016, pp. 3996–4003.

217. Jing Lu and Vincent Ng. "Event Coreference Resolution: A Survey of Two Decades of Research". In: *Proceedings of the Twenty-Seventh International Joint Conference on Artificial Intelligence*. California: International Joint Conferences on Artificial Intelligence Organization, July 2018, pp. 5479–5486. ISBN: 9780999241127. DOI: https://doi.org/10. 24963/.cai.2018/773. URL: https://www..cai.org/proceedings/2018/773.

218. R. Duncan Luce and John W. Tukey. "Simultaneous conjoint measurement: A new type of fundamental measurement". In: *Journal of Mathematical Psychology* 1.1 (Jan. 1964), pp. 1–27. ISSN: 00222496. DOI: https://doi.org/10.1016/0022-2496(64)90015-X. URL: https://linkinghub.elsevier.com/retrieve/pii/002224966490015X.

219. Luca Luceri, Silvia Giordano, and Emilio Ferrara. "Detecting Troll Behavior via Inverse Reinforcement Learning: A Case Study of Russian Trolls in the 2016 US Election". In: *Proceedings of the Fourteenth International AAAI Conference on Web and Social Media (ICWSM 2020)*. Association for the Advancement of ArtificialIntelligence, 2020, pp. 417–427. arXiv: 2001.10570. URL: http://arxiv.org/abs/2001.10570.

220. Brent MacGregor. *Live, direct, and biased? making television news in the satellite age*. Arnold, 1997.

221. Farzaneh Mahdisoltani, Joanna Biega, and Fabian M. Suchanek. "YAGO 3: A Knowledge Base from MultilingualWikipedias". In: *CIDR 2015, Seventh Biennial Conference on Innovative Data Systems Research*. www.cidrdb.org, 2015, pp. 1–11.

222. Oded Maimon and Lior Rokach. "Introduction to knowledge discovery and data mining". In: Data mining and knowledge discovery handbook. Springer, 2009, pp. 1–15.

223. Daniel deVassimon Manela et al. "Stereotype and Skew: Quantifying Gender Bias in Pretrained and Fine-tuned Language Models". In: (Jan. 2021). arXiv: 2101.09688. URL: http://arxiv.org/abs/2101.09688.

224. Christopher Manning et al. "The Stanford CoreNLP Natural Language Processing Toolkit". In: *Proceedings of 52nd Annual Meeting of the Association for Computational Linguistics: System Demonstrations*. Stroudsburg, PA, USA: Association for Computational Linguistics, 2014, pp. 55–60. ISBN: 9781941643006. DOI: https://doi.org/10.3115/v1/P14-5010. arXiv: arXiv:1011.1669v3. url: http://aclweb.org/anthology/P14-5010.

225. Christopher D Manning, Hinrich Schütze, et al. *Foundations of statistical natural language processing*. Vol. 999. MIT Press, 1999. ISBN: 9780262133609.

226. Christopher D. Manning, Prabhakar Raghavan, and Hinrich Schutze. *Introduction to Information Retrieval*. Cambridge: Cambridge University Press, 2008. ISBN: 9780511809071. DOI: https://doi.org/10.1017/CBO9780511809071. URL: http://ebooks.cambridge.org/ref/id/CBO9780511809071.

227. Jörg Matthes. "What's in a Frame? A Content Analysis of Media Framing Studies in the World's Leading Communication Journals, 1990-2005". In: *Journalism & Mass Communication Quarterly* 86.2 (June 2009), pp. 349–367. ISSN: 1077-6990. DOI: https://doi.org/10.1177/107769900908600206. URL: http://journals.sagepub.com/doi/10.1177/107769900908600206.

228. John McCarthy et al. "Assessing stability in the patterns of selection bias in newspaper coverage of protest during the transition from communism in Belarus". In: *Mobilization: An International Quarterly* 13.2 (2008), pp. 127–146.

229. John D McCarthy, Clark McPhail, and Jackie Smith. "Images of Protest: Dimensions of Selection Bias in Media Coverage of Washington Demonstrations, 1982 and 1991". In: *American Sociological Review* 61.3 (June 1996), p. 478. ISSN: 00031224. DOI: https://doi.org/10.2307/2096360. URL: http://www.jstor.org/stable/2096360?origin=crossref.

230. Meridith McGraw. Trump says he split with 'Mr. Tough Guy' Bolton over 'very big mistakes'. 2019. URL: https://abcnews.go.com/Politics/trumpsplit-mr-tough-guy-bolton-big-mistakes/story?id=65544651 (visited on 04/10/2021).

231. Margaret J McGregor et al. "Why don't more women report sexual assault to the police?" In: *Canadian Medical Association Journal* 162.5 (2000), pp. 659–660.

232. KathleenRMcKeown et al. "Tracking and summarizing news on a daily basis with Columbia's Newsblaster". In: *Proceedings of the second international conference on Human Language Technology Research*. 2002, pp. 280–285.

233. Merriam-Webster Inc. *The Merriam-Webster DictionaryNew Edition*. Springfield, MA, US, 2016, pp. 1–960. ISBN: 978-0877792956.

234. Philipp Meschenmoser et al. "Scraping Scientific Web Repositories: Challenges and Solutions for Automated Content Extraction". In: *D-Lib Magazine* 22.9/10 (Sept. 2016). ISSN: 1082-9873. DOI: https://doi.org/10.1045/september2016-meschenmoser. URL: http://www.dlib.org/dlib/september16/meschenmoser/09meschenmoser.html.

235. Norman Meuschke and Bela Gipp. "State-of-the-art in detecting academic plagiarism". In: *International Journal for Educational Integrity* 9.1 (June 2013), p. 50. ISSN: 1833-2595. DOI: https://doi.org/10.21913/.EI.v9i1.847. URL: https://ojs.unisa.edu.au/index.php/.EI/article/view/847.

236. Norman Meuschke et al. "HyPlag". In: *The 41st International ACM SIGIR Conference on Research & Development in Information Retrieval*. NewYork, NY, USA: ACM, June 2018, pp. 1321–1324. ISBN: 9781450356572. DOI: https://doi.org/10.1145/3209978.3210177. URL: https://dl.acm.org/doi/10.1145/3209978.3210177.

237. Joshua Meyrowitz. *No sense of place: The impact of electronic media on social behavior*. Oxford University Press, 1986.

238. Tomas Mikolov et al. "Efficient Estimation of Word Representations in Vector Space". In: *Proceedings of the 1st International Conference on Learning Representations (ICRL 2013)*. 2013, pp. 1–12. arXiv: 1301.3781. URL: http://arxiv.org/abs/1301.3781.

239. George A. Miller. "WordNet". In: *Communications of the ACM* 38.11 (Nov. 1995), pp. 39–41. ISSN: 0001-0782. DOI: https://doi.org/10.1145/219717.219748. URL: https://dl.acm.org/doi/10.1145/219717.219748.

240. Jim Miller. *A Critical Introduction to Syntax*. London, UK: Continuum International Publishing Group, 2011, pp. 1–288. ISBN: 978-0826497048.

241. M. Mark Miller. "Frame Mapping and Analysis of News Coverage of Contentious Issues". In: *Social Science Computer Review* 15.4 (Dec. 1997), pp. 367–378. ISSN: 0894-4393. DOI: https://doi.org/10.1177/089443939701500403. URL: http://journals.sagepub.com/doi/10.1177/089443939701500403.

242. Anne-Lyse Minard et al. "MEANTIME, the NewsReader Multilingual Event and Time Corpus". In: *Proceedings of the Tenth International Conference on Language Resources and Evaluation (LREC '16)*. Ed. by European Language Resources Association (ELRA). Portoroz, Slovenia, 2016, pp. 4417–4422. url: https://www.aclweb.org/anthology/L16-1699/.

243. Gilad Mishne. "Experiments with mood classification in blog posts". In: *Proceedings of ACM SIGIR 2005 Workshop on Stylistic Analysis of Text for Information Access* (2005).

244. Rishabh Misra. News Category Dataset. 2018. URL: https://www.kaggle.com/rmisra/news-category-dataset (visited on 02/19/2021).

245. Amy Mitchell et al. Political Polarization and Media Habits - From Fox News to Facebook, How Liberals and Conservatives Keep Up with Politics. Tech. rep. Pew Research Center, 2014, pp. 1–81. URL: http://www.pewresearch.org/pj_14-10-21_mediapolarization-08-2/.

246. Ryan Mitchell. *Web scraping with Python: collecting data from the modern web*. O'Reilly Media, Inc., 2015.

247. Saif Mohammad and Peter Turney. "Emotions Evoked by Common Words and Phrases: Using Mechanical Turk to Create an Emotion Lexicon". In: *Proceedings of the NAACL HLT 2010 Workshop on Computational Approaches to Analysis and Generation of Emotion in Text*. Los Angeles, CA: Association for Computational Linguistics, 2010, pp. 26–34.

248. Shunji Mori, Hirobumi Nishida, and Hiromitsu Yamada. *Optical character recognition*. John Wiley & Sons, Inc., 1999.

249. Karen Mossberger, Caroline J Tolbert, and Ramona S McNeal. *Digital citizenship: The Internet, society, and participation*. MIT Press, 2007. ISBN: 9780262134859. URL: https://mitpress.mit.edu/books/digital-citizenship.

250. Sendhil Mullainathan and Andrei Shleifer. "The market for news". In: *American Economic Review* (2005), pp. 1031–1053.
251. Sean A Munson and Paul Resnick. "Presenting diverse political opinions". In: *Proceedings of the 28th international conference on Human factors in computing systems -CHI '10*. ACM. NewYork, NewYork, USA: ACMPress, 2010, p. 1457. ISBN: 9781605589299. DOI: https://doi.org/10.1145/1753326.1753543. URL: http://portal.acm.org/citation.cfm?doid=1753326.1753543.
252. Sean A Munson, Daniel Xiaodan Zhou, and Paul Resnick. "Sidelines: An Algorithm for Increasing Diversity in News and Opinion Aggregators." In: *ICWSM*. 2009.
253. Sean A. Munson, Stephanie Y. Lee, and Paul Resnick. "Encouraging reading of diverse political viewpoints with a browser widget". In: *Proceedings of the 7th International Conference on Weblogs and Social Media, ICWSM 2013*. 2013.
254. Diana C Mutz. "Facilitating communication across lines of political difference: The role of mass media". In: *American Political Science Association*. Vol. 95. 01. Cambridge Univ Press. 2001, pp. 97–114.
255. David Nadeau and Satoshi Sekine. "A survey of named entity recognition and classification". In: *Lingvisticae Investigationes* 30.1 (Aug. 2007), pp. 3–26. ISSN: 0378-4169. DOI: https://doi.org/10.1075/li.30.1.03nad. URL: http://www.jbeplatform.com/content/journals/10.1075/li.30.1.03nad.
256. Sebastian Nagel. *Common Crawl: News Crawl*. 2016. URL: https://web.archive.org/web/20191118111519/https://commoncrawl.org/2016/10/news-dataset-available/ (visited on 03/24/2021).
257. Preslav Nakov et al. "SemEval-2013 Task 2: Sentiment Analysis in Twitter". In: *Second Joint Conference on Lexical and Computational Semantics (SEM), Volume 2: Proceedings of the Seventh International Workshop on Semantic Evaluation (SemEval 2013)*. Atlanta, GA, USA: Association for Computational Linguistics, 2013, pp. 312–320. URL: https://www.aclweb.org/anthology/S13-2052/.
258. Preslav Nakov et al. "SemEval-2016 Task 4: Sentiment Analysis in Twitter". In: *Proceedings of the 10th International Workshop on Semantic Evaluation (SemEval-2016)*. San Diego, CA, USA: Association for Computational Linguistics, 2016, pp. 1–18. DOI: https://doi.org/10.18653/v1/S16-1001.
259. Joseph Napolitan. *The election game and how to win it*. Doubleday, 1972.
260. Kimberly A Neuendorf. *The content analysis guidebook*. Sage Publications, 2016. ISBN: 9781412979474.
261. Nic Newman, David A L Levy, and Rasmus Kleis Nielsen. *Reuters Institute Digital News Report 2015*. Reuters Institute for the Study of Journalism, 2015. ISBN: 978-1907384134.
262. Nic Newman et al. *Reuters Institute Digital News Report 2020*. Reuters Institute for the Study of Journalism, 2020.
263. David Niven. Tilt? *The search for media bias*. Praeger, 2002. ISBN: 978–0275975777.
264. Jeppe Norregaard, Benjamin Horne, and Sibel Adali. "NELA-GT-2018: A large multi-labelled news dataset for the study of misinformation in news articles". In: *Proceedings of the Thirteenth International AAAI Conference on Web and Social Media*. AAAI Press, 2019, pp. 630–638.
265. ED O'Sullivan and SJ Schofield. "Cognitive bias in clinical medicine". In: *Journal of the Royal College of Physicians of Edinburgh* 48.3 (2018), pp. 225–232. ISSN: 14782715. DOI: https://doi.org/10.4997/JRCPE.2018.306. URL: https://www.rcpe.ac.uk/sites/default/files/jrcpe_48_3_osullivan.pdf.
266. Daniela Oelke, Benno Geißelmann, and Daniel A Keim. "Visual Analysis of Explicit Opinion and News Bias in German Soccer Articles". In: *Euro- Vis Workshop on Visual Analytics*. Vienna, Austria, 2012. DOI: 10.2312/PE/EuroVAST/EuroVA12/049-053. URL: https://doi.org/10.2312/PE/EuroVAST/EuroVA12/049-053.

267. Pamela E. Oliver and Gregory M. Maney. "Political Processes and Local Newspaper Coverage of Protest Events: From Selection Bias to Triadic Interactions". In: *American Journal of Sociology* 106.2 (Sept. 2000), pp. 463–505. ISSN: 0002-9602. DOI: https://doi.org/10.1086/316964. URL: http://www.journals.uchicago.edu/doi/10.1086/316964.

268. Oxford English. *Oxford English Dictionary*. Oxford: Oxford University Press, 2009.

269. Lawrence Page et al. *The PageRank citation ranking: bringing order to the web*. Tech. rep. 1999.

270. Georgios Paliouras et al. "PNS: A Personalized News Aggregator on the Web". In: *Intelligent interactive systems in knowledge-based environments*. Ed. by George A. Tsihrintzis and Maria Virvou. Berlin, Germany: Springer, 2008, pp. 175–197. ISBN: 978-3-540-77471-6. DOI: https://doi.org/10.1007/978-3-540-77471-6_10. URL: http://link.springer.com/10.1007/978-3-540-77471-6_10.

271. Zhongdang Pan and GeraldKosicki. "Framing analysis: An approach to news discourse". In: *Political Communication* 10.1 (1993), pp. 55–75. ISSN: 1058-4609. DOI: https://doi.org/10.1080/10584609.1993.9962963. URL: http://www.tandfonline.com/doi/abs/10.1080/10584609.1993.9962963.

272. Bo Pang and Lillian Lee. "Opinion mining and sentiment analysis". In: *Foundations and trends in information retrieval* 2.1-2 (2008), pp. 1–135. doi: https://doi.org/10.1561/1500000011.

273. Bo Pang and Lillian Lee. "Seeing stars". In: *Proceedings of the 43rd Annual Meeting on Association for Computational Linguistics - ACL '05*. Morristown, NJ, USA: Association for Computational Linguistics, 2005, pp. 115–124. DOI: https://doi.org/10.3115/1219840.1219855. URL: http://portal.acm.org/citation.cfm?doid=1219840.1219855.

274. Zizi Papacharissi and Maria de Fatima Oliveira. "News Frames Terrorism: A Comparative Analysis of Frames Employed in Terrorism Coverage in U.S. and U.K. Newspapers". In: *The International Journal of Press/Politics* 13.1 (Jan. 2008), pp. 52–74. ISSN: 1940-1612. DOI: https://doi.org/10.1177/1940161207312676. url: http://journals.sagepub.com/doi/10.1177/1940161207312676.

275. Souneil Park, KyungSoon Lee, and Junehwa Song. "Contrasting Opposing Views of News Articles on Contentious Issues". In: *Proceedings of the 49th Annual Meeting of the Association for Computational Linguistics: Human Language Technologies*. Portland, Oregon, USA: Association for Computational Linguistics, 2011, pp. 340–349. URL: https://www.aclweb.org/anthology/P11-1035.

276. Souneil Park et al. "NewsCube". In: *Proceedings of the 27th international conference on Human factors in computing systems - CHI 09*. New York, New York, USA: ACM Press, 2009, p. 443. ISBN: 9781605582467. DOI: https://doi.org/10.1145/1518701.1518772. URL: http://dl.acm.org/citation.cfm?doid=1518701.1518772.

277. Souneil Park et al. "NewsCube 2.0: An Exploratory Design of a Social News Website for Media Bias Mitigation". In: *Workshop on Social Recommender Systems*. 2011.

278. Souneil Park et al. "The politics of comments". In: *Proceedings of the ACM 2011 conference on Computer supported cooperative work - CSCW '11*. ACM. New York, New York, USA: ACM Press, 2011, p. 113. ISBN: 9781450305563. DOI: https://doi.org/10.1145/1958824.1958842. URL: http://portal.acm.org/citation.cfm?doid=1958824.1958842.

279. Kristen Parton et al. "Who, what, when, where, why? comparing multiple approaches to the cross-lingual 5W task". In: Proceedings of the Joint Conference of the 47th Annual Meeting of the ACL and the 4th International Joint Conference on Natural Language Processing of the AFNLP: Volume 1-Volume 1. Association for Computational Linguistics. 2009, pp. 423–431. url: https://www.aclweb.org/anthology/P09-1048.

280. Ajay Patel et al. "Magnitude: A Fast, Efficient Universal Vector Embedding Utility Package". In: *Proceedings of the 2018 Conference on EmpiricalMethods in Natural Language Processing: System Demonstrations*. Stroudsburg, PA, USA: Association for Computational Linguistics, 2018, pp. 120–126. DOI: https://doi.org/10.18653/v1/D18-2021. URL: http://aclweb.org/anthology/D18-2021.

281. Richard Paul and Linda Elder. *The Thinker's Guide for Conscientious Citizens on how to Detect Media Bias & Propaganda in National and World News.* Foundation Critical Thinking, 2004.

282. Richard M. Perloff. "A Three-Decade Retrospective on the Hostile Media Effect". In: Mass Communication and Society 18.6 (Nov. 2015), pp. 701–729. ISSN: 1520-5436. DOI: https://doi.org/10.1080/15205436.2015.1051234. URL: http://www.tandfonline.com/doi/full/10.1080/15205436.2015.1051234.

283. Cara Peters, Leigh Cellucci, and Daniel Kerrigan. "Improving the Hook in Case Writing". In: *Journal of Case Studies* 30 (2012), pp. 1–6.

284. Matthew Peters et al. "Deep Contextualized Word Representations". In: *Proceedings of the 2018 Conference of the North American Chapter of the Association for Computational Linguistics: Human Language Technologies, Volume 1 (Long Papers).* Stroudsburg, PA, USA: Association for Computational Linguistics, 2018, pp. 2227–2237. DOI: https://doi.org/10.18653/v1/N18-1202. URL: http://aclweb.org/anthology/N18-1202.

285. Pew Research Center. In *Changing U.S. Electorate, Race and Education Remain Stark Dividing Lines.* Tech. rep. 2020, pp. 1–41. URL: https://www.pewresearch.org/politics/2020/06/02/in-changing-u-s-electorate-race-andeducation-remain-stark-dividing-lines/.

286. Minh Hieu Phan and Philip O. Ogunbona. "Modelling Context and Syntactical Features for Aspect-based Sentiment Analysis". In: *Proceedings of the 58th Annual Meeting of the Association for Computational Linguistics. Stroudsburg, PA, USA: Association for Computational Linguistics, 2020,* pp. 3211–3220. DOI: https://doi.org/10.18653/v1/2020.acl-main.293. URL: https://www.aclweb.org/anthology/2020.acl-main.293.

287. Kathryn A Phillips, F. Reed Johnson, and Tara Maddala. "Measuring What People Value: A Comparison of "Attitude" and "Preference" Surveys". In: *Health Services Research* 37.6 (Dec. 2002), pp. 1659–1679. ISSN: 0017-9124. DOI: https://doi.org/10.1111/1475-6773.01116. URL: http://doi.wiley.com/10.1111/1475-6773.01116.

288. Evaggelia Pitoura et al. "On Measuring Bias in Online Information". In: *ACM SIGMOD Record* 46.4 (Feb. 2018), pp. 16–21. ISSN: 0163-5808. DOI: https://doi.org/10.1145/3186549.3186553. URL: https://dl.acm.org/doi/10.1145/3186549.3186553.

289. Maria Pontiki et al. "SemEval-2014 Task 4: Aspect Based Sentiment Analysis". In: *Proceedings of the 8th International Workshop on Semantic Evaluation (SemEval 2014).* Stroudsburg, PA, USA: Association for Computational Linguistics, 2014, pp. 27–35. DOI: https://doi.org/10.3115/v1/S14-2004. URL: http://aclweb.org/anthology/S14-2004.

290. Maria Pontiki et al. "SemEval-2015 Task 12: Aspect Based Sentiment Analysis". In: *Proceedings of the 9th International Workshop on Semantic Evaluation (SemEval 2015).* Stroudsburg, PA, USA: Association for Computational Linguistics, 2015, pp. 486–495. DOI: https://doi.org/10.18653/v1/S15-2082. URL: http://aclweb.org/anthology/S15-2082.

291. James Pustejovsky and Amber Stubbs. *Natural Language Annotation for Machine Learning: A guide to corpus-building for applications.* 1st ed. Sebastopol, CA, US: O'Reilly Media, Inc., 2012. ISBN: 9781449306663.

292. James Pustejovsky et al. "TimeML: Robust specification of event and temporal expressions in text". In: *New directions in question answering* 3 (2003), pp. 28–34.

293. Joaquin Quionero-Candela et al. *Dataset Shift in Machine Learning.* The MIT Press, 2009. ISBN: 9780262170055.

294. Dragomir R Radev et al. "Centroid-based summarization of multiple documents". In: *Information Processing & Management* 40.6 (2004), pp. 919–938.

295. Vikas C. Raykar et al. "Learning From Crowds". In: *The Journal of Machine Learning Research* 11 (2010), pp. 1297–1322. DOI: https://doi.org/10.5555/1756006.1859894.

296. Nornadiah Mohd Razali and BeeWah Yap. "Power comparisons of Shapiro-Wilk, Kolmogorov-Smirnov, Lilliefors and Anderson-Darling tests". In: *Journal of Statistical Modeling and Analytics* 2.1 (2011), pp. 21–33.

297. Marta Recasens, Cristian Danescu-Niculescu-Mizil, and Dan Jurafsky. "Linguistic Models for Analyzing and Detecting Biased Language". In: *Proceedings of the 51st Annual Meeting on Association for Computational Linguistics.* Sofia, BG: Association for Computational

Linguistics, 2013, pp. 1650–1659. ISBN: 9781937284503. URL: https://www.aclweb.org/anthology/P13-1162.pdf.

298. Marta Recasens, Eduard Hovy, and M. Antonia Marti. "A Typology of Near-Identity Relations for Coreference (NIDENT)". In: *Proceedings of the Seventh International Conference on Language Resources and Evaluation (LREC'10)*. Valletta, Malta: European Language Resources Association (ELRA), 2010, pp. 149–156. URL: http://www.lrec-conf.org/proceedings/lrec2010/pdf/160_Paper.pdf.

299. Marta Recasens, M. Antonia Marti, and Constantin Orasan. "Annotating Near-Identity from Coreference Disagreements". In: *Proceedings of the Eighth International Conference on Language Resources and Evaluation (LREC'12)*. Istanbul, Turkey: European Language Resources Association (ELRA), 2012, pp. 165–172. URL: https://www.aclweb.org/anthology/L12-1391/.

300. Alexander Rietzler et al. "Adapt or Get Left Behind: Domain Adaptation through BERT Language Model Finetuning for Aspect-Target Sentiment Classification". In: *Proceedings of the 12th Language Resources and Evaluation Conference*. European Language Resources Association, 2020. URL: https://www.aclweb.org/anthology/2020.lrec-1.607.

301. Frankie Robertson, Jarkko Lagus, and Kaisla Kajava. "A COVID-19 news coverage mood map of Europe". In: Proceedings of the EACL Hackashop on News Media Content Analysis and Automated Report Generation. Online: Association for Computational Linguistics, 2021, pp. 110–115. URL: https://www.aclweb.org/anthology/2021.hackashop-1.15.

302. JakobRogstadius et al. "An Assessment of Intrinsic and Extrinsic Motivation on Task Performance in Crowdsourcing Markets". In: *Proceedings of the Fifth International Conference on Weblogs and Social Media*. Ed. by Lada Adamic, Ricardo Baeza-Yates, and Scott Counts. Barcelona, Spain: AAAI Press, 2011, pp. 1–9. URL: https://www.aaai.org/ocs/index.php/ICWSM/ICWSM11/paper/view/2778.

303. Simcha Ronen and Oded Shenkar. "Attitudinal and Behavioral Dimensions". In: *Navigating Global Business*. Cambridge: Cambridge University Press, 2017, pp. 189–244. DOI: https://doi.org/10.1017/9781316107034.006. URL: http://ebooks.cambridge.org/ref/id/CBO9781316107034A037.

304. Shawn W Rosenberg et al. "The image and the vote: The effect of candidate presentation on voter preference". In: *American Journal of Political Science* 30.1 (1986), pp. 108–127. DOI: https://doi.org/10.2307/2111296.

305. Sara Rosenthal, Noura Farra, and Preslav Nakov. "SemEval-2017 Task 4: Sentiment Analysis in Twitter". In: *Proceedings of the 11th International Workshop on Semantic Evaluation (SemEval-2017)*. Vancouver, Canada: Association for Computational Linguistics, 2017, pp. 502–518. DOI: https://doi.org/10.18653/v1/s17-2088.

306. Barbara Rychalska et al. "Samsung Poland NLP Team at SemEval-2016 Task 1: Necessity for diversity; combining recursive autoencoders,WordNet and ensemble methods to measure semantic similarity." In: *Proceedings of the 10th International Workshop on Semantic Evaluation (SemEval-2016)*. Stroudsburg, PA, USA: Association for Computational Linguistics, 2016, pp. 602–608. ISBN: 9781941643952. DOI: https://doi.org/10.18653/v1/S16-1091. URL: http://aclweb.org/anthology/S16-1091.

307. Diego Saez-Trumper, Carlos Castillo, and Mounia Lalmas. "Social media news communities". In: *Proceedings of the 22nd ACM international conference on Conference on information & knowledge management - CIKM '13*. New York, New York, USA: ACM Press, 2013, pp. 1679–1684. ISBN: 9781450322638. DOI: https://doi.org/10.1145/2505515.2505623. URL: http://dl.acm.org/citation.cfm?doid=2505515.2505623.

308. Gerard Salton and Christopher Buckley. "Term-weighting approaches in automatic text retrieval". In: *Information processing and management* 24.5 (1988), pp. 513–523. DOI: https://doi.org/10.1016/0306-4573(88)90021-0.

309. Mark Sanderson. "Duplicate detection in the Reuters collection". In: *"Technical Report (TR-1997-5) of the Department of Computing Science at the University of Glasgow G12 8QQ, UK"* (1997).

310. Victor Sanh et al. "DistilBERT, a distilled version of BERT: smaller, faster, cheaper and lighter". In: *arXiv preprint arXiv: 1910.01108* (Oct. 2019). arXiv: 1910.01108. URL: http:// arxiv.org/abs/1910.01108.

311. Cicero Nogueira dos Santos and Maira Gatti. "Deep Convolutional Neural Networks for Sentiment Analysis of Short Texts". In: *Proceedings of COLING 2014, the 25th International Conference on Computational Linguistics: Technical Papers*. 2014, pp. 69–78. URL: https:// www.aclweb.org/anthology/C14-1008.

312. Frane Šarié et al. "Takelab: Systems for Measuring Semantic Text Similarity". In: *Proceedings of the First Joint Conference on Lexical and Computational Semantics-Volume 1: Proceedings of the main conference and the shared task, and Volume 2: Proceedings of the Sixth InternationalWorkshop on Semantic Evaluation*. Association for Computational Linguistics, 2012, pp. 441–448. URL: https://www.aclweb.org/anthology/S12-1060.

313. Raimund Schatz, Sebastian Egger, and Kathrin Masuch. "The Impact of Test Duration on User Fatigue and Reliability of Subjective Quality Ratings". In: *Journal of the Audio Engineering Society* 60 (2012), pp. 63–73. URL: http://www.aes.org/e-lib/browse.cfm?elib=16167.

314. Dietram A Scheufele. "Agenda-setting, priming, and framing revisited: Another look at cognitive effects of political communication". In: *Mass Communication & Society* 3.2-3 (2000), pp. 297–316. DOI: 10.1207/S15327825MCS0323_07. URL: https://doi.org/10.1207/ S15327825MCS0323_07.

315. Alexander von Schönburg. Wir hören zu viel auf Virologen! 2020. URL: https://www.bild. de/politik/inland/politik-inland/corona-skepsis-wirhoeren-zu-viel-auf-virologen-69738054. bild.html (visited on 02/19/2021).

316. Margrit Schreier. Qualitative content analysis in practice. SAGE Publications, 2012, pp. 1– 280. ISBN: 9781849205931.

317. Crisitina Segalin et al. "The Pictures We Like Are Our Image: Continuous Mapping of Favorite Pictures into Self-Assessed and Attributed Personality Traits". In: IEEE Transactions on Affective Computing 8.2 (Apr. 2017), pp. 268–285. ISSN: 1949-3045. DOI: https://doi.org/ 10.1109/TAFFC.2016.2516994. URL: http://ieeexplore.ieee.org/document/7378902/.

318. Anup Shah. Media Conglomerates, Mergers, Concentration of Ownership. 2009. URL: https://www.globalissues.org/article/159/media-conglomeratesmergers-concentration- of-ownership (visited on 02/19/2021).

319. Walid Shalaby, Wlodek Zadrozny, and Hongxia Jin. "Beyond word embeddings: learning entity and concept representations from large scale knowledge bases". In: Information Retrieval Journal (2018), pp. 1–18. DOI: s10791-018-9340-3. URL: https://doi.org/10.1007/ s10791-018-9340-3.

320. Shanya Sharma, Manan Dey, andKoustuv Sinha. "Evaluating Gender Bias in Natural Language Inference". In: reviewed for the International Conference on Learning Representations (ICLR 2021). 2021, pp. 1–11. URL: https://openreview.net/pdf?id=bnuU0PzXl0-.

321. Smriti Sharma et al. "News Event Extraction Using 5W1H Approach & Its Analysis". In: International Journal of Scientific & Engineering Research 4.5 (2013), pp. 2064–2068. URL: https://www..ser.org/onlineResearchPaperViewer.aspx?News-Event-Extraction-Using- 5W1HApproach-Its-Analysis.pdf.

322. Aaron D. Shaw, John J. Horton, and Daniel L. Chen. "Designing incentives for inexpert human raters". In: Proceedings of the ACM 2011 conference on Computer supported cooperative work - CSCW '11. New York, New York, USA: ACM Press, 2011, p. 275. ISBN: 9781450305563. DOI: https://doi.org/10.1145/1958824.1958865. URL: http://portal.acm.org/ citation.cfm?doid=1958824.1958865.

323. Peng Shi and Jimmy Lin. "Simple BERT Models for Relation Extraction and Semantic Role Labeling". In: arXiv preprint arXiv: 1904.05255 (Apr. 2019), pp. 1–6. arXiv: 1904.05255. URL: http://arxiv.org/abs/1904.05255.

324. Narayanan Shivakumar and Hector Garcia-Molina. "SCAM: A Copy Detection Mechanism for Digital Documents". In: In Proceedings of the Second Annual Conference on the Theory and Practice of Digital Libraries. 1995. URL: http://ilpubs.stanford.edu:8090/95/.

325. Alison Smith, Timothy Hawes, and Meredith Myers. "Hiérarchie: Interactive Visualization for Hierarchical Topic Models". In: *Proceedings of the Workshop on Interactive Language Learning, Visualization, and Interfaces*. Association for Computational Linguistics, 2014, pp. 71–78. ISBN: 9781941643150. DOI: https://doi.org/10.3115/v1/W14-3111.

326. Jackie Smith et al. "From Protest to Agenda Building: Description Bias in Media Coverage of Protest Events in Washington, D.C." In: *Social Forces* 79.4 (2001), pp. 1397–1423. URL: https://www.jstor.org/stable/2675477.

327. Norman Solomon. "Media Bias". In: *New Political Science* 24.2 (June 2002), pp. 293–297. ISSN: 0739-3148. DOI: https://doi.org/10.1080/073931402200145252. URL: http://www.tandfonline.com/doi/abs/10.1080/073931402200145252.

328. Samuel R Sommers et al. "Race and media coverage of Hurricane Katrina: Analysis, implications, and future research questions". In: *Analyses of Social Issues and Public Policy* 6.1 (2006), pp. 39–55. DOI: https://doi.org/10.1111/j.1530-2415.2006.00103.x.

329. Youwei Song et al. "Targeted Sentiment Classification with Attentional Encoder Network". In: *Artificial Neural Networks and Machine Learning - ICANN 2019: Text and Time Series*. Cham, US: Springer International Publishing, 2019, pp. 93–103. DOI: https://doi.org/10.1007/978-3-030-30490-4_9. URL: http://link.springer.com/10.1007/978-3-030-30490-4_9.

330. Robert Speer and Catherine Havasi. "Representing General Relational Knowledge in ConceptNet 5". In: *Proceedings of the Eight International Conference on Language Resources and Evaluation (LREC'12)*. Istanbul, Turkey: European Language Resources Association (ELRA), 2012. ISBN: 978-2-9517408-7-7. URL: http://www.lrec-conf.org/proceedings/lrec2012/pdf/1072_Paper.pdf.

331. Spiegel Online. Übertreibt Horst Seehofer seine Attacken? Das sagen die Medien. 2016. URL: http://www.spiegel.de/politik/deutschland/uebertreibthorst-seehofer-seine-attacken-das-sagen-die-medien-a-1076867.html (visited on 02/15/2021).

332. Timo Spinde et al. "Enabling News Consumers to View and Understand Biased News Coverage: A Study on the Perception and Visualization of Media Bias". In: *Proceedings of the ACM/IEEE Joint Conference on Digital Libraries in 2020*. New York, NY, USA: ACM, Aug. 2020, pp. 389–392. ISBN: 9781450375856. DOI: https://doi.org/10.1145/3383583.3398619. URL: https://dl.acm.org/doi/10.1145/3383583.3398619.

333. Andreas Spitz and Michael Gertz. "Breaking theNews: Extracting the Sparse CitationNetwork Backbone of OnlineNews Articles". In: *Proceedings of the 2015 IEEE/ACM International Conference on Advances in Social Networks Analysis and Mining 2015*. ACM. 2015, pp. 274–279. DOI: https://doi.org/10.1145/2808797.2809380.

334. Stanford NLP Group. *Stanford JavaNLP API Documentation*. 2021. URL: https://nlp.stanford.edu/nlp/javadoc/javanlp/ (visited on 02/15/2021).

335. Ralf Steinberger et al. "Large-scale news entity sentiment analysis". In: *RANLP 2017 - Recent Advances in Natural Language Processing Meet Deep Learning*. Incoma Ltd. Shoumen, Bulgaria, Nov. 2017, pp. 707–715. isbn: 9789544520496. DOI: https://doi.org/10.26615/978-954-452-049-6_091. URL: http://www.acl-bg.org/proceedings/2017/RANLP%202017/pdf/RANLP091.pdf.

336. Steve Stemler. "An overview of content analysis". In: *Practical assessment, research & evaluation* 7.17 (2001), pp. 137–146.

337. Guido H Stempel. "The prestige press meets the third-party challenge". In: *Journalism & Mass Communication Quarterly* 46.4 (1969), pp. 699–706. DOI: https://doi.org/10.1177/107769906904600402.

338. Guido H Stempel and John W Windhauser. "The prestige press revisited: coverage of the 1980 presidential campaign". In: *Journalism and Mass Communication Quarterly* 61.1 (1984), p. 49. DOI: https://doi.org/10.1177/107769908406100107.

339. James Glen Stovall. "Coverage of 1984 presidential campaign". In: *Journalism and Mass Communication Quarterly* 65.2 (1988), p. 443. DOI: https://doi.org/10.1177/107769908806500227.

340. James Glen Stovall. "The third-party challenge of 1980: News coverage of the presidential candidates". In: *Journalism and Mass Communication Quarterly* 62.2 (1985), p. 266. DOI: https://doi.org/10.1177/107769908506200206.

341. Carlo Strapparava and Rada Mihalcea. "Semeval-2007 task 14: Affective text". In: *Proceedings of the 4th International Workshop on Semantic Evaluations*. Association for Computational Linguistics. Prague, Czech Republic, 2007, pp. 70–74. URL: https://www.aclweb.org/anthology/S07-1013/.

342. Joseph D Straubhaar. Media Now: Communication Media in Information Age. Thomson Learning, 2000.

343. Jannik Strötgen and Michael Gertz. "Multilingual and cross-domain temporal tagging". In: *Language Resources and Evaluation* 47.2 (June 2013), pp. 269–298. ISSN: 1574-020X. DOI: https://doi.org/10.1007/s10579-012-9179-y. URL: http://link.springer.com/10.1007/s10579-012-9179-y.

344. Pero Subasic and Alison Huettner. "Affect analysis of text using fuzzy semantic typing". In: *IEEE Transactions on Fuzzy Systems* 9.4 (2001), pp. 483–496. ISSN: 10636706. DOI: https://doi.org/10.1109/91.940962. URL: http://ieeexplore.ieee.org/document/940962/.

345. Fabian M Suchanek, Gjergji Kasneci, and Gerhard Weikum. "Yago". In: *Proceedings of the 16th international conference on World Wide Web - WWW '07*. New York, New York, USA: ACM Press, 2007, p. 697. ISBN: 9781595936547. DOI: https://doi.org/10.1145/1242572.1242667. URL: http://portal.acm.org/citation.cfm?doid=1242572.1242667.

346. Chi Sun, Luyao Huang, and Xipeng Qiu. "Utilizing BERT for Aspect-Based Sentiment Analysisvia Constructing Auxiliary Sentence". In: *Proceedings of the 2019 Conference of the North*. Stroudsburg, PA, USA: Association for Computational Linguistics, 2019, pp. 380–385. DOI: https://doi.org/10.18653/v1/N19-1035. URL: http://aclweb.org/anthology/N19-1035.

347. Chi Sun et al. "How to Fine-Tune BERT for Text Classification?" In: *Chinese Computational Linguistics*. Cham, US: Springer International Publishing, 2019, pp. 194–206. DOI: https://doi.org/10.1007/978-3-030-32381-3_16. URL: http://link.springer.com/10.1007/978-3-030-32381-3_16.

348. S Shyam Sundar. "Exploring receivers' criteria for perception of print and online news". In: *Journalism & Mass Communication Quarterly* 76.2 (1999), pp. 373–386. DOI: https://doi.org/10.1177/107769909907600213.

349. Ralph Sundberg and Erik Melander. "Introducing the UCDP Georeferenced Event Dataset". In: *Journal of Peace Research* 50.4 (July 2013), pp. 523–532. issn: 0022-3433. DOI: https://doi.org/10.1177/0022343313484347. URL: http://journals.sagepub.com/doi/10.1177/0022343313484347.

350. Beth M. Sundheim. "Overview of the fourth message understanding evaluation and conference". In: *Proceedings of the 4th conference on Message understanding - MUC4 '92*. Vol. 298. Morristown, NJ, USA: Association for Computational Linguistics, 1992, p. 3. ISBN: 1558602739. DOI: https://doi.org/10.3115/1072064.1072066. URL: http://dl.acm.org/citation.cfm?id=1072066.

351. Cass R Sunstein. *Echo Chambers: Bush v. Gore, Impeachment, and Beyond*. Princeton University Press Princeton, 2001.

352. Cass R Sunstein. "The law of group polarization". In: *Journal of political philosophy* 10.2 (2002), pp. 175–195. URL: https://papers.ssrn.com/sol3/papers.cfm?abstract_id=199668.

353. Christian Szegedy et al. "Rethinking the Inception Architecture for Computer Vision". In: *Proceedings of the IEEE conference on computer vision and pattern recognition*. IEEE, 2016, pp. 2818–2826. DOI: https://doi.org/10.1109/CVPR.2016.308. arXiv: 1512.00567. URL: http://arxiv.org/abs/1512.00567.

354. Tagesschau.de. Mehr als 37.000 Infizierte in Deutschland. 2020. URL: https://web.archive.org/web/20210121145909/https://www.tagesschau.de/newsticker/liveblog-coronavirus-131.html (visited on 04/13/2021).

355. Hristo Tanev, Jakub Piskorski, and Martin Atkinson. "Real-Time News Event Extraction for Global Crisis Monitoring". In: *Natural Language and Information Systems*. Vol. 5039 LNCS. Berlin, Heidelberg: Springer Berlin Heidelberg, 2008, pp. 207–218. ISBN: 3540698574. DOI:

https://doi.org/10.1007/978-3-540-69858-6_21. URL: http://link.springer.com/10.1007/978-3-540-69858-6_21.

356. Amanda Taub. How Liberals Got Lost on the Story of Missing Children at the Border. 2018. URL: https://web.archive.org/web/20191120151037/https://www.nytimes.com/2018/05/31/upshot/liberals-immigration-childrenborder-misinformation.html (visited on 02/18/2021).

357. Yla R. Tausczik and James W. Pennebaker. "The Psychological Meaning of Words: LIWC and Computerized Text Analysis Methods". In: *Journal of Language and Social Psychology* 29.1 (Mar. 2010), pp. 24–54. ISSN: 0261-927X. DOI: https://doi.org/10.1177/0261927X09351676. URL: http://journals.sagepub.com/doi/10.1177/0261927X09351676.

358. The Media Insight Project. The Personal News Cycle: How Americans Get Their News. Tech. rep. 2014. URL: https://www.americanpressinstitute.org/publications/reports/survey-research/personal-news-cycle/.

359. Erik F. Tjong Kim Sang and Fien De Meulder. "Introduction to the CoNLL-2003 Shared Task: Language-Independent Named Entity Recognition". In: *Proceedings of the Seventh Conference on Natural Language Learning at HLT-NAACL 2003*. 2003, pp. 142–147. URL: https://www.aclweb.org/anthology/W03-0419.

360. Trägerverein des Deutschen Presserats e.V. *Ethische Standards für den Journalismus*. 2021. URL: https://www.presserat.de/pressekodex.html (visited on 03/05/2021).

361. Manos Tsagkias, Maarten De Rijke, and Wouter Weerkamp. "Linking online news and social media". In: *Proceedings of the fourth ACM international conference on Web search and data mining*. ACM. 2011, pp. 565–574. DOI: https://doi.org/10.1145/1935826.1935906.

362. Larry Tye. *The father of spin: Edward L. Bernays and the birth of public relations*. Macmillan, 2002.

363. University of Michigan. News Bias Explored - The art of reading the news. 2014. URL: http://umich.edu/~newsbias/ (visited on 02/01/2021).

364. University of Michigan Library. "Fake News," Lies and Propaganda: How to Sort Fact from Fiction: Where do news sources fall on the political bias spectrum? 2021. URL: http://guides.lib.umich.edu/c.php?g=637508&p=4462444 (visited on 02/15/2021).

365. Christine D Urban. *Examining Our Credibility: Perspectives of the Public and the Press*. Asne Foundation, 1999, pp. 1–108.

366. Mojtaba Vaismoradi, Hannele Turunen, and Terese Bondas. "Content analysis and thematic analysis: Implications for conducting a qualitative descriptive study". In: *Nursing & Health Sciences* 15.3 (Sept. 2013), pp. 398–405. ISSN: 14410745. DOI: https://doi.org/10.1111/nhs.12048. URL: http://doi.wiley.com/10.1111/nhs.12048.

367. Robert P Vallone, Lee Ross, and Mark R Lepper. "The hostile media phenomenon: biased perception and perceptions of media bias in coverage of the Beirut massacre." In: *Journal of personality and social psychology* 49.3 (1985), p. 577. DOI: https://doi.org/10.1037//0022-3514.49.3.577.

368. Baldwin Van Gorp. "Strategies to take subjectivity out of framing analysis". In: *Doing news framing analysis: Empirical and theoretical perspectives* (2010), pp. 100–125. URL: https://www.taylorfrancis.com/chapters/edit/10.4324/9780203864463-11/strategies-take-subjectivity-framing-analysisbaldwin-van-gorp.

369. Baldwin Van Gorp. "Where is the frame? Victims and intruders in theBelgian press coverage of the asylum issue". In: *European Journal of Communication* 20.4 (2005), pp. 484–507. DOI: https://doi.org/10.1177/0267323105058253.

370. Athanasios Voulodimos et al. "Deep Learning for Computer Vision: A Brief Review". In: *Computational Intelligence and Neuroscience 2018* (2018), pp. 1–13. ISSN: 1687-5265. DOI: https://doi.org/10.1155/2018/7068349. URL: https://www.hindawi.com/journals/cin/2018/7068349/.

371. Paul Waldman and James Devitt. "Newspaper Photographs and the 1996 Presidential Election: The Question of Bias". In: *Journal of Mass Communication* 75.2 (1998), pp. 302–311. ISSN: 10776990. DOI: https://doi.org/10.1177/107769909807500206.

372. Emily Wall, John Stasko, and Alex Endert. "Toward a Design Space for Mitigating Cognitive Bias in Vis". In: 2019 IEEE Visualization Conference (VIS). IEEE, Oct. 2019, pp. 111–

115. ISBN: 978-1-7281-4941-7. DOI: https://doi.org/10.1109/VISUAL.2019.8933611. URL: https://ieeexplore.ieee.org/document/8933611/.

373. Alex Wang et al. "GLUE: A Multi-Task Benchmark and Analysis Platform for Natural Language Understanding". In: *Proceedings of the 2018 EMNLP Workshop BlackboxNLP: Analyzing and Interpreting Neural Networks for NLP*. Stroudsburg, PA, USA: Association for Computational Linguistics, 2018, pp. 353–355. DOI: https://doi.org/10.18653/v1/W18-5446. URL: http://aclweb.org/anthology/W18-5446.

374. WeiWang, Dongyan Zhao, and DongWang. "Chinese news event 5w1h elements extraction using semantic role labeling". In: *Proceedings of the 2010 Third International Symposium on Information Processing*. IEEE. 2010, pp. 484–489. DOI: https://doi.org/10.1109/ISIP.2010.112.

375. Wayne Wanta, Guy Golan, and Cheolhan Lee. "Agenda setting and international news: Media influence on public perceptions of foreign nations". In: *Journalism & Mass Communication Quarterly* 81.2 (2004), pp. 364–377. DOI: https://doi.org/10.1177/107769900408100209.

376. Ralph Weischedel et al. OntoNotes 5.0. 2013. URL: URL https://web.archive.org/web/20190705173013, https://catalog.ldc.upenn.edu/LDC2013T19 (visited on 02/19/2021).

377. David Manning White. "The "Gate Keeper": A Case Study in the Selection of News". In: Journalism Bulletin 27.4 (1950), pp. 383–390. ISSN: 0197-2448. DOI: https://doi.org/10.1177/107769905002700403. URL: http://journals.sagepub.com/doi/10.1177/107769905002700403.

378. Michael L.Wick et al. "A unified approach for schema matching, coreference and canonicalization". In: *Proceeding of the 14th ACM SIGKDD international conference on Knowledge discovery and data mining - KDD 08*. New York, New York, USA: ACM Press, 2008, p. 722. ISBN: 9781605581934. doi: https://doi.org/10.1145/1401890.1401977. URL: http://dl.acm.org/citation.cfm?doid=1401890.1401977.

379. Philine Widmer, Elliott Ash, and Sergio Galletta. "Media Slant is Contagious". In: *SSRN Electronic Journal* (2020), pp. 1–42. ISSN: 1556-5068. DOI: https://doi.org/10.2139/ssrn.3712218. URL: https://www.ssrn.com/abstract=3712218.

380. Ben FM W.nen et al. "Discrete-choice experiments versus rating scale exercises to evaluate the importance of attributes". In: *Expert Review of Pharmacoeconomics & Outcomes Research* 15.4 (July 2015), pp. 721–728. ISSN: 1473-7167. DOI: https://doi.org/10.1586/14737167.2015.1033406. URL: http://www.tandfonline.com/doi/full/10.1586/14737167.2015.1033406.

381. Wikinews. Main Page -Wikinews, The free news source you can write. 2021. url: https://en.wikinews.org/w/index.php?title=Main_Page&oldid=4045302 (visited on 02/15/2021).

382. Alden Williams. "Unbiased Study of Television News Bias". In: *Journal of Communication* 25.4 (Dec. 1975), pp. 190–199. ISSN: 0021-9916. DOI: https://doi.org/10.1111/j.1460-2466.1975.tb00656.x. URL: https://academic.oup.com/joc/article/25/4/190-199/4553978.

383. Theresa Wilson, Janyce Wiebe, and Paul Hoffmann. "Recognizing contextual polarity in phrase-level sentiment analysis". In: Proceedings of the conference on Human Language Technology and Empirical Methods in Natural Language Processing - HLT '05. Morristown, NJ, USA: Association for Computational Linguistics, 2005, pp. 347–354. DOI: https://doi.org/10.3115/1220575.1220619. URL: http://portal.acm.org/citation.cfm?doid=1220575.1220619.

384. Stephane Wolton. "Are Biased Media Bad for Democracy?" In: *American Journal of Political Science* 63.3 (July 2019), pp. 548–562. ISSN: 0092-5853. doi: https://doi.org/10.1111/ajps.12424. URL: https://onlinelibrary.wiley.com/doi/abs/10.1111/ajps.12424.

385. Robert F. Woolson, Judy A. Bean, and Patricio B. Rojas. "Sample Size for Case-Control Studies Using Cochran's Statistic". In: *Biometrics* 42.4 (Dec. 1986), p. 927. ISSN: 0006341X. DOI: https://doi.org/10.2307/2530706. URL: https://www.jstor.org/stable/2530706?origin=crossref.

386. Yonghui Wu et al. "Google's Neural Machine Translation System: Bridging the Gap between Human and Machine Translation". In: Preprint (Sept. 2016). arXiv: 1609.08144. URL: http://arxiv.org/abs/1609.08144.

387. Adrian deWynter and Daniel J. Perry. "Optimal Subarchitecture Extraction For BERT". In: (Oct. 2020). arXiv: 2010.10499. URL: http://arxiv.org/abs/2010.10499.

388. Wei Xiang and Bang Wang. "A Survey of Event Extraction From Text". In: *IEEE Access* 7 (2019), pp. 173111–173137. ISSN: 2169-3536. DOI: https://doi.org/10.1109/ACCESS.2019. 2956831. URL: https://ieeexplore.ieee.org/document/8918013/.

389. Liyan Xu and Jinho D. Choi. "Revealing the Myth of Higher-Order Inference in Coreference Resolution". In: *Proceedings of the 2020 Conference on Empirical Methods in Natural Language Processing (EMNLP)*. Stroudsburg, PA, USA: Association for Computational Linguistics, 2020, pp. 8527–8533. doi: https://doi.org/10.18653/v1/2020.emnlp-main.686. URL: https://www.aclweb.org/anthology/2020.emnlp-main.686.

390. Liyan Xu and Jinho D. Choi. "Revealing the Myth of Higher-Order Inference in Coreference Resolution". In: *Proceedings of the 2020 Conference on Empirical Methods in Natural Language Processing (EMNLP)*. Stroudsburg, PA, USA: Association for Computational Linguistics, 2020, pp. 8527–8533. DOI: https://doi.org/10.18653/v1/2020.emnlp-main.686. URL: https://www.aclweb.org/anthology/2020.emnlp-main.686.

391. Vikas Yadav and Steven Bethard. "A Survey on Recent Advances in Named Entity Recognition from Deep Learning models". In: *Proceedings of the 27th International Conference on Computational Linguistics. Santa Fe, New Mexico, USA: Association for Computational Linguistics*, 2018, pp. 2145–2158. URL: https://www.aclweb.org/anthology/C18-1182.

392. Sibel Yaman, Dilek Hakkani-Tur, and Gokhan Tur. "Combining semantic and syntactic information sources for 5-w question answering". In: *INTERSPEECH*. 2009, pp. 2707–2710. URL: https://www.isca-speech.org/archive/archive_papers/interspeech_2009/papers/i09_2707.pdf.

393. Sibel Yaman et al. "Classification-based strategies for combining multiple 5-w question answering systems". In: *Tenth Annual Conference of the International Speech Communication Association*. Brighton,UK: International Speech Communication Association, 2009, pp. 1–4. URL: https://www.iscaspeech.org/archive/archive_papers/interspeech_2009/papers/i09_2703.pdf.

394. Heng Yang. LC-ABSA. 2020. URL: https://github.com/yangheng95/LCABSA/blob/master/models/lc_apc/lcf_bert.py (visited on 04/15/2021).

395. JungHwan Yang et al. "Why Are "Others" So Polarized? Perceived Political Polarization and Media Use in 10 Countries". In: *Journal of Computer-Mediated Communication* 21.5 (Sept. 2016), pp. 349–367. ISSN: 10836101. doi: https://doi.org/10.1111/jcc4.12166. URL: https://academic.oup.com/jcmc/article/21/5/349-367/4161799.

396. Lucas Ou-Yang. *Newspaper: Article scraping & curation*. 2021. URL: http://newspaper.readthedocs.io/en/latest/ (visited on 02/17/2021).

397. Yiming Yang, Tom Pierce, and Jaime Carbonell. "A study of retrospective and on-line event detection". In: *Proceedings of the 21st annual international ACM SIGIR conference on Research and development in information retrieval - SIGIR '98*. New York, New York, USA: ACM Press, 1998, pp. 28–36. ISBN: 1581130155. DOI: https://doi.org/10.1145/290941. 290953. URL: http://portal.acm.org/citation.cfm?doid=290941.290953.

398. Da Yin, Tao Meng, and Kai-Wei Chang. "SentiBERT: A Transferable Transformer-Based Architecture for Compositional Sentiment Semantics". In: *Proceedings of the 58th Annual Meeting of the Association for Computational Linguistics*. Stroudsburg, PA, USA: Association for Computational Linguistics, 2020, pp. 3695–3706. DOI: https://doi.org/10.18653/v1/2020.acl-main.341. url: https://www.aclweb.org/anthology/2020.acl-main.341.

399. John Zaller. *The nature and origins of mass opinion*. Cambridge university press, 1992. DOI: https://doi.org/10.1017/CBO9780511818691.

400. Biqing Zeng et al. "LCF: A Local Context Focus Mechanism for Aspect-Based Sentiment Classification". In: *Applied Sciences* 9.16 (Aug. 2019), pp. 1–22. ISSN: 2076-3417. DOI: https://doi.org/10.3390/app9163389. URL: https://www.mdpi.com/2076-3417/9/16/3389.

401. Bowen Zhang et al. "Enhancing Cross-target Stance Detection with Transferable Semantic-Emotion Knowledge". In: *Proceedings of the 58th Annual Meeting of the Association for Computational Linguistics*. Stroudsburg, PA, USA: Association for Computational Linguis-

tics, 2020, pp. 3188–3197. DOI: https://doi.org/10.18653/v1/2020.acl-main.291. URL: https://www.aclweb.org/anthology/2020.acl-main.291.

402. Lei Zhang, ShuaiWang, and Bing Liu. "Deep learning for sentiment analysis: A survey". In: *Wiley Interdisciplinary Reviews: Data Mining and Knowledge Discovery* 8.4 (July 2018). ISSN: 1942-4787. DOI: https://doi.org/10.1002/widm.1253. URL: https://onlinelibrary.wiley.com/doi/abs/10.1002/widm.1253.

403. Anastasia Zhukova et al. "XCoref: Cross-document Coreference Resolution in the Wild". In: *Proceedings of the iConference*. IEEE, 2022, pp. 1–10.

404. Jochen Zöllner et al. "Optimizing small BERTs trained for German NER". In: (Apr. 2021). arXiv: 2104.11559. URL: http://arxiv.org/abs/2104.11559.

405. Sven Meyer Zu Eissen and Benno Stein. "Intrinsic plagiarism detection". In: *European Conference on Information Retrieval. Springer*. 2006, pp. 565–569. URL: https://link.springer.com/chapter/10.1007/11735106_66.

Printed in the United States
by Baker & Taylor Publisher Services